Green Transformation in the Digital Age

Weike Zhang • Ming Zeng • Ao Yu

Green Transformation in the Digital Age

Financial Innovation and Technological Empowerment in China's Carbon Neutrality Process

palgrave
macmillan

Weike Zhang
School of Public Administration
Sichuan University
Chengdu, China

Ao Yu
School of Economics
Sichuan University
Chengdu, China

Ming Zeng
School of Economics
Xihua University
Chengdu, China

ISBN 978-981-95-9120-6 ISBN 978-981-95-9121-3 (eBook)
https://doi.org/10.1007/978-981-95-9121-3

This Palgrave Macmillan imprint is published by the registered company Springer Nature Singapore Pte Ltd.
The registered company address is: 152 Beach Road, #21-01/04 Gateway East, Singapore 189721, Singapore

Preface

We are living through a great recalibration. The steady, predictable march of progress has given way to a more volatile era, defined by a convergence of ecological, technological, and financial forces. The climate is no longer a mere backdrop to human affairs but an active, unstable participant. In response, two powerful engines are being retooled: the engine of technology, which is weaving a digital nervous system through our economies, and the engine of finance, which is being rewired to value and direct capital flows based on a new, carbon-aware logic. It is at the confluence of these three transformative currents, namely the imperative of a green transition, the ascent of the digital age, and the re-architecting of finance, that this book finds its purpose. Our inquiry is focused on China, a civilization-scale entity engaged in one of history's most ambitious balancing acts. Its challenge is not merely to sustain developmental momentum while pivoting toward carbon neutrality, but to orchestrate a symphony between its technological prowess and its financial imagination. The official targets of "peaking by 2030, neutral by 2060" are more than policy goals; they are a live experiment in whether a modern economy can be fundamentally rewired and its capital and technologies can be redirected by policy to fuel a greener existence.

This book is the product of years of observing this immense project not as distant spectators but as researchers embedded in its complexities. We have traced the pathways of green credit as it filters from central bank directives to the loan portfolios of provincial branches, and we have documented how algorithms are being deployed to optimize everything from citywide power grids to the kiln temperatures of a single cement plant. A central question emerged: Is this fusion of finance and technology creating a new, more sophisticated toolkit for sustainability, or is it merely optimizing an outdated system?

The chapters that follow are our attempt to answer this. They are built not on theoretical postulates, but on empirical ground—data from hundreds of cities and thousands of enterprises. We examine how artificial intelligence, often discussed in abstract terms, concretely reshapes urban energy metabolism and corporate governance. We investigate whether financial innovation, such as green credit, truly rewards exploratory green innovation or incentivizes low-risk compliance. The Chinese context, with its unique blend of state capacity and market dynamism, offers a powerful laboratory. Its successes provide a blueprint, and its struggles reveal universal challenges.

This book maps a contested frontier—a terrain where the future is being forged not by a single force, but by the dynamic, often fraught, interplay of two. On the one side lies the formidable potential of technology, with its algorithmic precision and intelligent systems. On the other stands the transformative power of financial innovation, actively rewriting the rules of capital allocation. Their convergence defines the new economy of our time. We trace not just the promise of this convergence but also its frictions: the clash between digital foresight and institutional inertia, and the struggle to align the immediate logic of financial returns with the long-term imperative of decarbonization. We offer this work as an intervention into that struggle, for building a viable future requires us to master the tools reshaping our world.

Chengdu, China W. Zhang
Chengdu, China M. Zeng
Chengdu, China A. Yu
October 2025

Acknowledgments

We are profoundly thankful to our co-authors—Qiwei Zhao, Hongxia Fan, Qian Luo, and Yufeng Zhang—whose dedicated work forms the empirical backbone of this book. Specifically, Chap. 1 was completed by Weike Zhang, Ming Zeng, and Ao Yu. Chapter 2 was completed by Ming Zeng and Weike Zhang. Chapter 3 was completed by Weike Zhang, Hongxia Fan, and Ming Zeng. Chapter 4 was completed by Weike Zhang, Hongxia Fan, and Qiwei Zhao. Chapter 5 was completed by Ming Zeng and Weike Zhang. Chapter 6 was completed by Weike Zhang and Ming Zeng. Chapter 7 was completed by Weike Zhang, Qian Luo, Yufeng Zhang, and Ao Yu. Chapter 8 was completed by Weike Zhang, Ming Zeng, and Ao Yu. We wish to thank Lei Zhou for his efforts in helping to prepare the manuscript. Finally, we extend our sincere appreciation to the exceptional team at Springer Nature, particularly our editors Jacob Dreyer and Raju Heggidi, for their skillful guidance and steadfast support throughout the publication process. We are also very grateful for the support of the National Natural Science Foundation of China (Grant no. 72004150) and Sichuan Natural Science Foundation (Grant no. 2025NSFSC1934).

Competing Interests The authors have no competing interests to declare that are relevant to the content of this manuscript.

Contents

About the Authors

Weike Zhang is a professor at the School of Public Administration of Sichuan University, China. His research focus is on public policy and technology policy. Over the past five years, he has served as the principal investigator of 15 longitudinal projects, including national science and technology major projects, national social science later-stage funding projects, and national natural science foundation projects. He has published over 60 academic papers in SSCI/CSSCI journals. Among them, 11 papers have been selected as ESI highly cited papers. He serves as a guest editor or editorial board member for several journals and also acts as a reviewer for over 140 SSCI/SCI/CSSCI journals.

Ming Zeng is an assistant professor at the School of Economics, Xihua University, China. His research specializes in sustainable finance and technology-driven environmental governance, with a focus on low-carbon transitions in emerging markets. Dr. Zeng has published extensively in top-tier SSCI journals (*Energy Economics* and *Sustainable Cities and Society*). He currently leads a Ministry of Education-funded project and co-investigates National Social Science Fund projects. Dr. Zeng teaches financial theory and cutting-edge topics to graduate students and serves as a reviewer for leading journals like *Cities and Energy Economics.*

Ao Yu is a professor at the School of Economics, Sichuan University, China. His research focuses on the digital economy, science and technology economy, state-owned and private enterprises, and judicial systems. Professor Yu has published over 40 papers in core journals, including those indexed in SSCI and CSSCI, and has authored or co-edited four academic books. He has hosted more than 40 research projects funded by the National Social Science Fund of China, the Ministry of Education, and Sichuan Province. He teaches undergraduate and graduate courses, such as political economy, new institutional economics, and enterprise economics. In addition to his academic work, Professor Yu actively engages in policy advising and public service. He serves as a government consultant at both provincial and municipal levels, and as an external director appointed by Chengdu SASAC.

List of Figures

List of Tables

CHAPTER 1

Introduction

1.1 Background

1.1.1 Emissions at a Peak: The Critical Decade to Secure Carbon Neutrality

Climate change has emerged as an urgent global challenge, prompting unprecedented commitments by countries to achieve carbon neutrality (Zeng & Zhang, 2024; Zhang & Zeng, 2024; Zhang et al., 2023a, b, 2025). According to the International Energy Agency (IEA), global energy-related CO_2 emissions reached approximately 37.8 gigatonnes of carbon dioxide equivalent ($GtCO_2e$) in 2024, a historical high, which represents a year-on-year increase of about 0.8%.[1] The continued rise in greenhouse gas emissions has led to a sustained increase in global average temperatures, more frequent extreme weather events, and heightened climate risks, drawing intense international attention and accelerating carbon-neutral transition efforts (Zeng & Zhang, 2024; Zhang et al., 2025). The adoption of the Paris Agreement in 2015 marked a milestone in the global determination to address climate change, setting the goal of limiting global warming to well below 2 °C, with efforts to restrict it to 1.5 °C. However, under current policy trajectories, global temperatures are projected to rise well above 2 °C, highlighting a significant gap between

[1] IEA (2025), Global Energy Review 2025, IEA, Paris.

W. Zhang et al., *Green Transformation in the Digital Age*,
https://doi.org/10.1007/978-981-95-9121-3_1

ambition and reality. The latest Emissions Gap Report by the United Nations Environment Programme (UNEP) warns that without a substantial intensification of mitigation actions within this decade, the 1.5 °C target will become unattainable. To achieve this goal, global emissions by 2030 must be reduced by an additional 45% compared with current pledges.[2] Achieving carbon neutrality has thus become one of the most pressing challenges facing human society today.

Building on this global context, countries are accelerating their green transformation through carbon neutrality pledges (Zhang & Zeng, 2024; Zhang et al., 2023a). As of November 2023, about 145 countries had announced or were considering net-zero targets by the mid-twenty-first century, collectively accounting for nearly 90% of global emissions.[3] Major economies such as the European Union, the United States, and Japan have pledged to reach carbon neutrality by 2050, while emerging economies such as China and India have set their respective peaking and neutrality timelines. These commitments cover the bulk of global greenhouse gas emissions, indicating a broad international consensus on the low-carbon transition (Zhang et al., 2023a). Nevertheless, a significant gap persists between ambition and action: many countries have yet to introduce sufficiently robust policies to deliver on their nationally determined contributions (NDCs), leaving the pace of global mitigation far short of the Paris Agreement objectives. Global emissions have not declined but have continued to rise. In 2023, global greenhouse gas emissions reached a record high of 57.1 billion tons of carbon dioxide equivalent, an increase of 1.3% compared to 2022, and fully surpassing pre-pandemic levels.[4] This is mainly attributed to the growth in the energy, transportation, and industrial sectors (Zhang & Zeng, 2024). Carbon dioxide concentrations and global temperatures have also reached unprecedented levels, triggering more frequent extreme weather events worldwide. Nowadays, every fraction of a degree increase in temperature brings sharply rising risks, ranging

[2] United Nations Environment Programme (2024). Executive summary. In Emissions Gap Report 2024: No more hot air … please! With a massive gap between rhetoric and reality, countries draft new climate commitments. Nairobi.

[3] https://climateactiontracker.org/global/cat-net-zero-target-evaluations/#:~:text=As%20of%20November%202023%2C%20around,of%20global%20greenhouse%20gas%20emissions

[4] United Nations Environment Programme (2024). Executive summary. In Emissions Gap Report 2024: No more hot air … please! With a massive gap between rhetoric and reality, countries draft new climate commitments. Nairobi.

from deadly heatwaves and floods to food and water insecurity. UNEP estimates that, without additional measures, global warming could reach approximately 2.9 °C by 2100.[5] Bridging this "commitment–emission" gap will require leapfrogging in policy, technological, and financial actions over the coming years.

China, as the world's largest emitter of greenhouse gases, plays a pivotal role in the global transition toward carbon neutrality (Zhang et al., 2023b, 2025). In 2020, the Chinese government formally announced its "dual carbon" targets of peaking carbon emissions before 2030 and achieving carbon neutrality before 2060, which demonstrates its resolve to shoulder the responsibilities of a major power and to pursue a green, low-carbon development path. Following this high-level commitment, China moved swiftly to establish a comprehensive climate policy architecture to realize the 2030/2060 goals, with the core being the "1+N" policy framework for carbon peaking and carbon neutrality introduced in 2021. In this framework, the "1" refers to the overarching guideline (top-level design document) for achieving these goals, while the "N" denotes a series of sectoral action plans and sub-strategies tailored to specific industries and thematic areas. In recent years, China has released a national action plan for carbon peaking alongside multiple sector-specific policies covering energy, industry, transport, buildings, and agriculture, collectively forming a roadmap toward 2030 and beyond. These documents set out concrete objectives—such as controlling coal consumption, improving energy efficiency (EE), and expanding the share of nonfossil fuels—effectively integrating climate goals into the overall economic system. Over the past five years, China's CO_2 emissions intensity per unit of gross domestic product (GDP) has continued to decline. Despite this progress, however, due to the country's vast economic scale and an energy structure still dominated by fossil fuels, total emissions remain high. In 2024, China's energy-related CO_2 emissions were approximately 12.6 billion tonnes, accounting for about one-third of the global total.[6] Coal continues to dominate China's energy mix, despite steady growth in the share of nonfossil energy. In 2023, total energy consumption reached 5.72 billion tonnes of standard coal equivalent, with coal accounting for around 55% (Song, 2025).

[5] https://www.carbonbrief.org/unep-new-climate-pledges-need-quantum-leap-in-ambition-to-deliver-paris-goals/#:~:text=The%20report%20warns%20that%20both,in%202030

[6] IEA (2025), Global Energy Review 2025, IEA, Paris.

1.1.2 Action in the Digital Age: Collective Forces Driving Green Transformation

Cities and enterprises stand as the pivotal actors in the green transformation, translating global ambitions into tangible action through digital innovation, investment, and collaboration at both the theoretical and practical levels (Zeng & Zhang, 2024; Zhang et al., 2023a, 2025). Over half of the world's population resides in urban areas, a proportion projected to increase to approximately two-thirds by 2050.[7] Cities are the core arenas of energy consumption and carbon emissions, with urban activities accounting for an estimated 75–80% of final energy use and around 70% of CO_2 emissions globally.[8] Urbanization and economic growth continue to drive rising energy demand from buildings, transport, and industry, creating structural barriers to emissions reduction (Zeng & Zhang, 2024). Chinese cities have adopted carbon peaking action plans and smart city initiatives to accelerate their green transformation. For example, Chinese cities are actively implementing supportive policies to promote the development of renewable energy through a comprehensive framework encompassing financial incentives, legal regulations, technical assistance, and construction standards (Lee & Wang, 2024). Meanwhile, China has fully embraced digital technologies to advance smart city development, systematically enhancing urban digital governance, environmental quality, and management efficiency, thereby strengthening its capacity for green production (Zhang et al., 2025; Jiang et al., 2021). In addition, cities across China are proactively and systematically advancing the comprehensive development of cutting-edge digital infrastructure (DI)—such as 5G networks, Internet of Things (IoT) systems, and smart data platforms—with the dedicated aim of fostering sustainable, green, and low-carbon urban development through enhanced energy efficiency, intelligent transportation, and resilient ecological management (Zhang et al., 2023a; Tang & Yang, 2023).

Enterprises, as microeconomic actors, are the direct implementers of low-carbon transitions (Zhang & Zeng, 2024; Zhang et al., 2023b). Emissions reductions in high-energy-consuming sectors—such as steel,

[7] https://www.unep.org/topics/sustainable-development-goals/why-do-sustainable-development-goals-matter/goal-11-0#:~:text=By%202050%2C%2070%20per%20cent,consumption%20and%2075%20per

[8] https://blogs.worldbank.org/en/sustainablecities/cutting-global-carbon-emissions-where-do-cities-stand#:~:text=stand%3F%20blogs,huge%20quantities%20of%20fossil

cement, and chemicals—and in large firms directly influence the achievement of macro targets. In recent years, the rise of environmental, social, and governance (ESG) principles in global capital markets has prompted firms to place greater emphasis on their carbon footprints and sustainability performance. Hundreds of prominent global companies have announced net-zero visions, with over 500 obtaining Science Based Targets initiative (SBTi) net-zero validation.[9] And a 2024 study of the world's 2000 largest companies found that about 16% are on pathways consistent with their 2050 net-zero targets.[10] Chincsc enterprises are increasingly recognizing their environmental and social responsibilities, having transitioned from merely disclosing ESG information to proactively engaging in substantive practices (Gao et al., 2025; Wang et al., 2024); they are now incorporating sustainable and low-carbon principles at the core of their enterprise strategies and daily operations, and are committed to pursuing technological innovation and enhanced quality while simultaneously reducing resource consumption (Zhang & Zeng, 2024; Li et al., 2023; Tan & Zhu, 2022).

1.1.3 Finance as a Catalyst: Mobilizing Strategic Capital for Green Transformation

Global momentum is building to harness financial innovation to drive deep decarbonization. Green finance (GF)—spanning instruments such as green bonds, green loans, and transition finance—has expanded rapidly worldwide, mobilizing unprecedented capital for low-carbon projects. The global market for sustainable finance reached about USD 4.2 trillion in 2023 and is projected to grow more than sixfold by 2033 (Lou, 2025). This reflects a sweeping shift among investors and policymakers to align financial flows with climate goals. China has emerged as a key player in this green finance surge: backed by ambitious policies, its green credit and bond issuances grew at an average annual rate of about 30% from 2019 to 2023 (Lou, 2025), funding renewable energy, clean transport, and other green infrastructure. Such financial innovation is recognized as pivotal for achieving net-zero pathways, as it reallocates capital towards sustainable industries and incentivizes enterprises to adopt cleaner production (Zhang

[9] https://sciencebasedtargets.org/blog/500-companies-net-zero-ambition#:~:text=Net,looks%20set%20to%20grow%20quickly

[10] Accenture. Destination Net Zero[EB/OL]. (2024-10-24).

et al., 2023b). By offering favorable terms for green projects and imposing higher costs on carbon-intensive activities, green finance mechanisms help align economic incentives with emissions reduction objectives. Notably, global institutions are scaling up support: the World Bank Group delivered a record $42.6 billion in climate finance in 2024 alone to back mitigation and resilience efforts.[11] Yet a large financing gap persists, especially in developing economies: the IMF warns that emerging countries will need to mobilize trillions of dollars in additional investment beyond their domestic means to hit net-zero by 2050,[12] highlighting the indispensable role of both public and private capital.

China is proactively utilizing financial innovation as a catalyst to advance its green transformation. In 2021, it launched a national Emissions Trading System (ETS), initially covering the power sector (about 4 billion tonnes of CO_2). The ETS has since become the largest of its kind in the world, with cumulative transactions reaching 465 million tonnes and a total value of nearly RMB 27 billion as of July 2024.[13] Although carbon prices remain relatively low and the system is still being extended to additional high-emission sectors such as cement, steel, and aluminum smelting, it has already established a market-based mechanism to drive long-term emission reductions. In parallel, China continues to position green finance as a key driver in achieving its carbon neutrality goals. The People's Bank of China (PBOC) and other financial regulators have developed one of the world's largest green finance markets, with outstanding green loans exceeding RMB 40 trillion (approximately USD 5.6 trillion) by the first quarter of 2025.[14] The PBOC has also introduced innovative instruments such as the Carbon Emission Reduction Facility (CERF), which offers relending to banks at an annual interest rate of 1.75%, covering up to 60% of the principal amount of eligible green loans.[15] Supported by policy tools and financial innovation, China's renewable energy and clean technology sectors have grown rapidly.

[11] https://www.worldbank.org/en/news/press-release/2024/09/19/climate-finance-fiscal-year-2024-snapshot#:~:text=WASHINGTON%2C%20September%2019%2C%202024%20,compared%20to%20the%20previous%20year

[12] https://www.imf.org/en/Topics/climate-change/COP29#:~:text=To%20hit%20net%20zero%20by,investment%2C%20or%20FDI%2C%20is%20key

[13] https://www.gov.cn/yaowen/liebiao/202407/content_6963787.htm

[14] https://www.gov.cn/lianbo/bumen/202506/content_7026533.htm

[15] http://www.pbc.gov.cn/redianzhuanti/118742/4357176/4357209/4384182/index.html

1.1.4 *Technology as a Vanguard: Engineering Practical Tools for Green Transformation*

In parallel, technological empowerment—especially digital technologies and AI—is transforming how emissions are monitored and reduced (Zhang et al., 2025). AI-powered solutions and digital infrastructure are boosting energy and carbon efficiency across energy systems: smart grids and IoT sensors optimize electricity distribution and integrate renewables in real time, while machine learning models help detect leaks and inefficiencies in industrial processes.[16] For example, AI-driven analytics can improve fuel and logistics efficiency in transport by 5–10% and cut building energy use by around 10% through intelligent controls.[17] At scale, adopting existing AI applications in energy, manufacturing, and cities could reduce global CO_2 emissions by an estimated 1.4 $GtCO_2e$ by 2035, substantially outweighing the carbon footprint of data centers powering these tools.[18]

Driven by these enabling conditions, China now ranks first globally in installed renewable power capacity. By the end of 2024, renewables—including hydropower, wind, solar, and biomass—accounted for around 56% of total installed power capacity, surpassing the national target of 1200 gigawatts (GW) six years ahead of schedule.[19] More broadly, China's clean technology sector—dominated by the so-called new three industries of solar photovoltaic modules, lithium batteries, and electric vehicles—expanded in 2024 at roughly three times the annual GDP growth rate. These industries recorded sales and investment totaling RMB 13.6 trillion (about USD 1.9 trillion), resulting in clean energy technologies accounting for over 10% of total economic output for the first time and contributing roughly one-quarter of national GDP growth, overtaking real estate as a primary growth driver.[20] China now supplies over 80% of global photovoltaic modules production, 70% of key raw materials for power batteries, and about 40% of global electric vehicles output.[21] This dual dynamic—rapid clean energy expansion alongside continued coal reliance—highlights

[16] IEA (2025), Energy and AI, IEA, Paris.

[17] IEA (2025), Energy and AI, IEA, Paris.

[18] IEA (2025), Energy and AI, IEA, Paris.

[19] https://climateactiontracker.org/countries/china/#:~:text=As%20in%20previous%20years%2C%20our,underscoring%20the%20need%20for%20rapid

[20] https://world.gmw.cn/2025-08/01/content_38193615.htm

[21] https://world.gmw.cn/2025-08/01/content_38193615.htm

the central challenge of balancing energy security with decarbonization in achieving the dual-carbon goals.

In China's push for carbon neutrality, these twin engines of financial innovation and technological empowerment are increasingly intertwined. Financial innovation is catalyzing the deployment of AI and digital systems (from urban smart grids to enterprise carbon management platforms) that drive down carbon intensity at city and enterprise levels (Zeng & Zhang, 2024). Concurrently, the measurable gains from technological empowerment-enabled efficiency and cleaner production strengthen the rationale for further green investments, creating a virtuous cycle of decarbonization (Zhang et al., 2025). As an example, China inaugurated its national carbon emissions trading system in 2021, now the world's largest carbon market covering over 4.5 billion tons of CO_2 annually.[22] Despite this progress, challenges remain. In the financial realm, standardization gaps and divergent interests may undermine the effectiveness of green finance initiatives (Lou, 2025), highlighting the need for clearer taxonomies and policy coordination. On the technology side, the rapid growth of AI and digital infrastructure raises concerns about its own environmental footprint—energy-hungry data centers and electronic waste, for instance, can offset some gains.[23] Ensuring that digital solutions truly deliver net emissions cuts will require greening the tech sector and mitigating rebound effects. Nonetheless, the strategic integration of financial innovation and technological empowerment stands out as a cornerstone of global and Chinese efforts to achieve carbon neutrality, offering scalable pathways to lower carbon intensity while fostering sustainable economic growth.

1.2 Conceptual Framework and Definitions

1.2.1 Carbon Neutrality

Carbon neutrality refers to a state in which, over a defined period (typically one year), carbon dioxide emissions from human activities are balanced by

[22] https://english.www.gov.cn/statecouncil/ministries/202107/17/content_WS60f210bfc6d0df57f98dd212.html#:~:text=The%20market%20currently%20includes%20over,Ministry%20of%20Ecology%20and%20Environment

[23] https://www.unep.org/news-and-stories/story/ai-has-environmental-problem-heres-what-world-can-do-about#:~:text=But%20when%20it%20comes%20to,warming%20greenhouse%20gases

human-led removal of carbon dioxide, resulting in net-zero emissions (Zhang & Zeng, 2024; Zhang et al., 2023a). To clarify this concept, two key points should be emphasized. First, in a narrow sense, "carbon" in carbon neutrality specifically means carbon dioxide. Broadly, it may also encompass all greenhouse gases. This monograph adopts the narrow definition. Second, carbon neutrality pertains only to anthropogenic emissions and removals, excluding natural carbon cycles. It does not imply the complete cessation of human-related CO_2 emissions, but rather requires that such emissions be offset through human-driven measures such as afforestation and artificial carbon capture, thereby neutralizing the net impact of human activities on the climate system.

In 2020, China pledged at the United Nations General Assembly to peak carbon dioxide emissions by 2030 and achieve carbon neutrality by 2060.[24] To fulfill this commitment, China has established a "1+N" policy framework and is implementing multidimensional measures across sectors such as industry, energy, transportation, and technology innovation.[25] These efforts aim to transition from a high-carbon development model to a sustainable, low-carbon paradigm. Thus, carbon neutrality—with net-zero CO_2 emissions as its ultimate objective—entails a systemic transformation encompassing energy, economy, technology, and finance.

1.2.2 Green Transformation

Green transformation is a fundamental process of change aimed at achieving sustainable development of human society by reshaping areas such as energy structures, industrial composition, social culture, and governance models, moving beyond inefficient development patterns reliant on natural resource inputs, promoting the decoupling of socioeconomic development from resource consumption and environmental pollution, and ultimately achieving harmonious coexistence between humans and nature (Zhang & Zeng, 2024; Zeng & Zhang, 2024). The concept can be understood through four key aspects. First, green transformation is a systematic, comprehensive, and directional societal shift involving profound changes in development philosophy, growth drivers, and modes of development—it is not merely minor environmental regulation. Second, it represents a harmonious integration of environmental protection and development; it

[24] https://www.qstheory.cn/yaowen/2020-09/22/c_1126527766.htm
[25] https://www.gov.cn/xinwen/2021-07/14/content_5624831.htm

does not seek to curb economic growth or reduce quality of life, but rather pursues higher energy efficiency and resource conservation within a framework of high-quality economic development. Third, green transformation is a long-term and dynamic process—not an overnight achievement. It addresses not a binary "yes or no" question, but rather the directional shift of the entire economy and society toward ecological harmony. Finally, although green transformation is a broad concept that is difficult to measure precisely, it can be monitored using common indicators such as energy efficiency, industrial structure, and technological innovation. In this monograph, metrics such as carbon emission intensity (CEI), ESG, and EE are used to characterize the green transformation.

EE generally refers to the ability to obtain the same or higher levels of service, output, or functionality while consuming less energy (Zeng & Zhang, 2024; Zhang et al., 2025). The IEA defines EE as "using less energy to provide the same service or achieve the same result," emphasizing the role of technological, behavioral, and managerial improvements in reducing energy demand without compromising performance.[26] In practical terms, EE can be measured as the ratio of useful output (e.g., goods produced, heating provided, kilometers traveled) to the energy input required, and improvements in EE reflect either reduced energy consumption for a given output or increased output per unit of energy consumed. In the context of the green transformation, EE is widely recognized as the "first fuel" for reducing greenhouse gas emissions, as identified by the IEA and the Intergovernmental Panel on Climate Change (IPCC).[27] Enhancing EE delivers multiple co-benefits: it reduces carbon emissions and other pollutants, lowers energy costs for households and businesses, and mitigates the need for new energy infrastructure investment. Empirical studies show that EE improvements have been a major driver of decoupling economic growth from energy consumption in many countries, with significant contributions to lowering CEI (Rosenow et al., 2017). In this book, EE is treated solely as an outcome of low-carbon transformation. Higher EE directly reduces CEI by improving the carbon productivity of economic activities, with AI playing a role in driving such improvements.

[26] IEA. (2022). Energy Efficiency 2022. International Energy Agency.

[27] IPCC. (2022). Climate Change 2022: Mitigation of Climate Change. Cambridge University Press.

CEI is a key indicator for assessing the carbon efficiency of economic activity, typically defined as the amount of CO_2 emitted per unit of economic output (Zhang et al., 2023a). More specifically, CEI is often expressed as CO_2 emissions per unit of GDP, representing the CO_2 released for each unit of GDP produced (Wang & Su, 2020). This metric captures the relationship between economic growth and carbon emissions: a decline in CEI generally signals industrial upgrading, improved EE, or a shift toward cleaner energy sources. For countries and regions that adopt carbon intensity targets—rather than absolute caps—reducing CEI serves as a primary pathway to achieving emission-reduction commitments while sustaining economic expansion. In practice, CEI is influenced by multiple factors, including the energy structure (share of fossil vs. nonfossil fuels), technological innovation, industrial composition, and the effectiveness of environmental governance. Lowering CEI requires not only improvements in EE and the adoption of renewable energy but also structural adjustments to production and consumption patterns. In this book, CEI is employed as a key analytical metric for assessing low-carbon transitions at urban levels.

ESG has become a prominent concept in the global investment arena, used to evaluate the overall sustainability performance of enterprises (Friede et al., 2015). It is not only a key reference in investment decision-making but also a core framework for assessing enterprise social responsibility and long-term value creation (Friede et al., 2015). ESG encompasses three core factors for assessing the sustainability and social impact of investments or companies: environmental, social, and governance. Unlike traditional investment analysis focused solely on financial returns, the ESG framework emphasizes integrating nonfinancial indicators related to sustainable development into decision-making, encouraging companies to generate profits while assuming environmental and social responsibilities. In recent years, growing awareness of climate risks and social equity has driven rapid growth in ESG investing, alongside the emergence of sustainable finance products such as ESG funds and sustainability-linked bonds. Research shows that companies with high ESG performance tend to exhibit better risk management and financial stability in the long run (Broadstock et al., 2021), attracting substantial capital to firms with higher ESG ratings. For corporations, proactive ESG strategies can enhance brand reputation, reduce compliance and operational risks, and improve competitiveness. In the carbon neutrality transition, the environmental component of ESG is especially critical, with investors and regulators increasingly focused on enterprise carbon disclosures, reduction targets, and green technology innovations. This book considers ESG as an important framework for evaluating enterprise sustainability performance, with particular attention to how AI can influence and enhance ESG outcomes in enterprises. The analysis focuses on the pathways through which AI-powered solutions can improve environmental performance, strengthen governance, and foster socially responsible practices, thereby supporting broader sustainability and low-carbon transition goals.

1.2.3 Financial Innovation

In this monograph, financial innovation specifically refers to innovations within the context of green transformation. It addresses the inadequacy of funding for green development under traditional financial systems by introducing new financial instruments, optimizing market mechanisms, and reforming policy frameworks—thereby guiding social capital toward green initiatives and facilitating sustainability transitions (Zeng & Zhang, 2024; Zhang et al., 2023b). Green transformation projects typically involve large upfront investments, long return periods, and significant risk uncertainties, which often lead to financing challenges. Through innovative tools such as green bonds, green credit, and carbon finance, along with new market mechanisms like carbon emission trading systems and supportive financial policies, the leveraging and resource allocation functions of finance can be fully utilized to mobilize societal funds and accelerate green transformation. This monograph adopts a green finance perspective to explore the concept of financial innovation.

GF generally refers to financial activities and institutional arrangements designed to support environmentally sustainable development (Zeng & Zhang, 2024). The World Economic Forum defines GF as any financial activity explicitly designed to deliver better environmental outcomes, including a range of loans, debt instruments, and investments that support green projects or minimize the environmental impact of traditional projects.[28] It encompasses traditional instruments such as green credit, green bonds, and green funds, as well as emerging domains including climate finance, sustainable investment, and carbon finance. Its essence lies in

[28] https://www.weforum.org/stories/2020/11/what-is-green-finance/#:~:text=,19

integrating environmental factors into investment and financing decisions, thereby directing capital flows toward sectors with ecological benefits—such as renewable energy, energy conservation, environmental protection, and clean transportation—thereby achieving a "win–win" of economic growth and environmental improvement (Zhang et al., 2019; Zeng & Zhang, 2024). In practice, GF not only mobilizes capital for low-carbon infrastructure but also drives innovation by encouraging enterprises to invest in cleaner technologies and sustainable business models. It can lower transition costs, reduce financing risks through mechanisms such as credit enhancement and interest subsidies, and attract broader social capital into green sectors (Hu et al., 2023; Zhang et al., 2023b). In this book, GF is regarded as a key driver for achieving carbon neutrality: through innovative financial mechanisms, it can lower the cost of low-carbon transitions, facilitate capital allocation to green industries, and provide financing support for emission-reduction projects at both enterprise and city levels.

1.2.4 Technological Empowerment

In this monograph, technological empowerment refers specifically to the enablement provided by technology within the context of green transformation. It entails the intelligent upgrading of conventional production technologies, business processes, and industrial chains through digital means—centered on technologies such as artificial intelligence (AI)—to comprehensively enhance low-carbon efficiency, optimize resource utilization, and reduce pollution emissions, thereby accelerating the green transformation (Zhang & Zeng, 2024; Zhang et al., 2023a, 2025). Grounded in the concept that science and technology constitute a primary productive force, technological empowerment supplies practical tools and methodologies to advance sustainability. This includes not only purely green and low-carbon technologies but also innovations spanning the entire production process. With a focus on digital infrastructure and artificial intelligence, this monograph examines the role and mechanisms of technological empowerment.

DI refers to the foundational hardware, software, and network systems that underpin the functioning of the digital economy (Zhang et al., 2023a). According to the International Data Center Authority (IDCA), DI includes physical and virtual systems such as data centers, broadband networks, and cloud platforms, which together constitute the "digital base" enabling the operation of the digital economy and public digital services.[29] Typical components include communication networks, computing infrastructure, the IoT, and smart terminal devices. High-quality DI construction is widely recognized as a catalyst for economic growth, innovative capacity, and social connectivity. Within the "dual carbon" context, DI plays a dual role. On the one hand, technologies such as IoT, cloud computing, and big data analytics, supported by robust DI, enable real-time monitoring, predictive control, and optimization of energy use in cities, industries, and supply chains, thereby reducing waste and improving overall efficiency. On the other hand, DI itself must undergo a green transformation through measures such as energy-efficient retrofitting of data centers, low-carbon construction of network facilities, and increased reliance on renewable-powered operations to mitigate the carbon footprint of digital expansion. In this book, technological empowerment is examined with a particular focus on DI, analyzing how its development and optimization can accelerate the green transformation.

AI has been one of the fastest-growing fields in digital technology in recent years (Zhang et al., 2025). IBM defines AI as the application of technology to enable computers and machines to simulate human learning, understanding, problem-solving, decision-making, creativity, and autonomy.[30] In other words, AI systems can perceive their environment, process information, learn from data, and take action to complete intelligent tasks that were previously limited to humans. Common AI technologies include machine learning, deep learning, natural language processing, and computer vision, which are widely applied in speech recognition, image analysis, autonomous driving, and intelligent recommendation systems. Nevertheless, most current applications fall under narrow AI, focused on specific tasks, while artificial general intelligence—possessing humanlike general capabilities—remains at a conceptual or early exploratory stage. In addressing climate change and promoting the green transformation, AI holds considerable promise: it can optimize energy systems, enhance industrial process efficiency, and improve building and transportation management, thereby reducing

[29] https://www.idc-a.org/government-support/digital-infrastructure

[30] https://www.ibm.com/think/topics/artificial-intelligence?utm_source=ts2.tech#:~:text=Artificial%20intelligence%20,decision%20making%2C%20creativity%20and%20autonomy#:~:text=What%20is%20AI%3F

resource waste and carbon emissions (Rolnick et al., 2022). AI can also be deployed in carbon emissions monitoring, climate risk forecasting, and environmental data analytics, providing scientific evidence to inform policymaking and investment decisions (Alshawi, 2022). At the same time, AI development must take into account its own energy use and ethical considerations, such as the electricity consumption of large data centers and the transparency and fairness of AI decision-making. This book considers AI an important component of technological empowerment, focusing on its potential applications in urban governance and enterprise low-carbon transitions.

1.3 Outline of the Book

From the global to the Chinese context, and from the urban to the enterprise level, the transition toward carbon neutrality is marked by both urgency and formidable challenges. Against this broad backdrop, the present volume seeks to construct a pioneering analytical framework that examines China's low-carbon transformation through the synergistic interaction of financial innovation—encompassing GF, digital inclusive finance, and green credit policy instruments—and technological empowerment, centered on AI. The book is organized into two principal parts, focusing respectively on the urban and enterprise dimensions, and concludes with a set of targeted policy recommendations. Each part is grounded in three empirical studies, systematically delineating the pathways through which technological empowerment and financial innovation shape sustainable development. Figure 1.1 illustrates the analytical framework of this study.

1.3.1 Urban-Level Analysis

Part I (Chaps. 2, 3, and 4*)* focuses on the mechanisms through which AI, GF, and DI shape urban EE and CEI governance. By examining these three dimensions—technology, finance, and infrastructure—this section elucidates the multifaceted drivers and interactive dynamics of low-carbon transitions at the urban scale, thereby advancing our understanding of sustainable development in the macro context.

Chapter 2 links financial resource allocation with technological diffusion to investigate the catalytic role of GF in advancing AI investment, promoting EE, and fostering low-carbon urban development. Drawing on

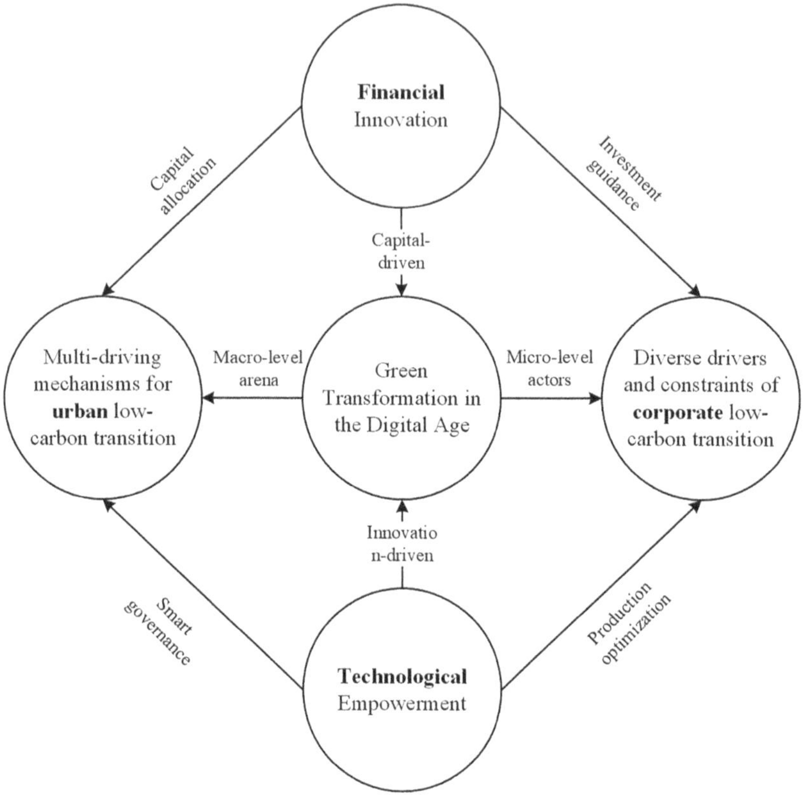

Fig. 1.1 Analytical framework

data from 282 Chinese cities between 2014 and 2019, the results show that GF significantly promotes AI adoption. Heterogeneity analysis suggests that this effect is more salient in eastern and central cities, non–resource-dependent cities, smaller cities with populations below five million, and cities with relatively low initial levels of AI. Further evidence demonstrates that GF not only enhances EE but also generates positive spatial spillover effects on urban EE.

Chapter 3 is grounded in the interactive logic between urban economies and energy systems and employs data from 282 Chinese cities spanning 2006 to 2019 to assess the impact of AI on urban EE. The findings indicate that AI significantly enhances urban EE. Further analysis reveals heterogeneity in this relationship across different city types: the positive effect of AI on EE is particularly pronounced in resource-dependent cities, cities in the eastern and central regions, those located north of the Qinling–Huaihe Line, as well as in megacities and super-large cities. Moreover, the study demonstrates a positive spatial spillover effect, whereby improvements in EE within a focal city are reinforced by the diffusion of AI applications from neighboring cities.

Chapter 4 explores the relationship between DI and environmental governance by examining the influence of DI on urban CEI, based on data from 275 prefecture-level cities during 2011–2019. The results reveal that a one-standard-deviation increase in DI corresponds to a 1.9% reduction in CEI. The heterogeneity analysis highlights that the effect of DI is particularly significant in cities north of the Qinling–Huaihe Line, east of the Hu Huanyong Line, as well as in eastern, central, and resource-dependent cities. Mechanism analysis further indicates that DI reduces CEI by promoting industrial upgrading, decreasing resource dependence, and stimulating green innovation. In addition, the study uncovers a negative spatial spillover effect, suggesting that the expansion of DI in one city can indirectly contribute to reductions in CEI in neighboring cities.

1.3.2 Enterprise-Level Analysis

Part II (Chaps. 5, 6, and 7) centers on how enterprises, as microeconomic actors, respond and adapt to the low-carbon transition, with particular emphasis on the pathways through which AI enhances firms' sustainable development performance and EE. By focusing on firms' internal processes of technology adoption, policy responses, and external institutional environments, this section reveals the diverse drivers and constraints shaping low-carbon transitions at the micro level, and underscores the synergy and complementarity between technological empowerment and policy guidance in facilitating enterprise green transformation.

Chapter 5 examines the integration of AI into the ESG framework, analyzing how technological adoption reshapes firms' ESG performance and how these effects vary across institutional and market environments. Drawing on data from 1270 listed firms in 187 Chinese cities between 2006 and 2019, the study finds that AI significantly improves enterprise ESG outcomes. The positive impact is especially salient for firms located in eastern and central cities, resource-based cities, high-tech industries, and state-owned enterprises (SOEs). Mechanism analysis suggests that AI enhances ESG performance by increasing analyst attention and accelerating digital transformation. Furthermore, higher levels of marketization and better business environments amplify the effectiveness of AI in improving ESG performance.

Chapter 6 investigates the effect of AI on firms' energy intensity (EI) using data from China's listed manufacturing firms between 2011 and 2019. The results indicate that AI adoption significantly reduces EI. The effect is particularly pronounced in energy-dependent firms, nonlabor-intensive firms, and SOEs. Mechanism analysis reveals that AI reduces EI primarily by fostering technological innovation and advancing digital transformation. The study also highlights that the business cycle, industry concentration, and environmental regulations condition the extent to which AI mitigates EI.

Chapter 7 incorporates policy instruments into the enterprise innovation framework by analyzing the impact of green credit policies on exploratory innovation, based on data from Chinese listed firms between 2007 and 2017. The findings demonstrate that green credit policies significantly promote exploratory innovation, with especially strong effects for SOEs, capital-intensive industries, and firms located in central China. Mechanism analysis further indicates that these policies stimulate exploratory innovation by effectively increasing R&D investment. Finally, while higher financing constraints typically exert adverse effects, the influence of green credit policies on exploratory innovation becomes more pronounced as the degree of financing constraint intensifies.

1.3.3 Policy Recommendations

Part III (Chap. 8) presents the policy recommendations derived from the empirical results of the preceding six chapters. This section aims to advance China's green transformation in support of its carbon neutrality goals. Given that facilitating such a transition is a complex undertaking requiring coordinated action, this study synthesizes policy recommendations across six interdependent dimensions: policy, finance, technology, markets, infrastructure, and multi-stakeholder engagement. Together, these elements form a coherent and mutually reinforcing framework. Specifically, policy sets the regulatory foundation; finance and technology supply critical implementation tools; markets and infrastructure provide the operational arena; and stakeholder collaboration serves as the driving force that integrates and activates the entire system.

As for policies, it is recommended to implement low-carbon-oriented policies, formulate differentiated financial support policies, and forge integrated finance and technology policy for green transformation. As for finance, it is recommended to scale up the provision of diversified green finance, develop innovative green financial instruments, and secure efficient and safe green capital allocation. As for technologies, it is recommended to promote low-carbon technology development, deepen the application of AI to improve energy efficiency, and systematize technology-driven decarbonization in all social domains.

As for markets, it is recommended to stimulate market-driven green innovation with strategic incentives, regulate market mechanisms to ensure fair green competition, and establish modern market regulatory frameworks. As for infrastructure, it is recommended to catalyze strategic investment in digital infrastructure, tailor digital infrastructure to local contexts, and integrate low-carbon into comprehensive urban planning. As for stakeholder collaboration, it is recommended to strengthen multilevel government coordination, advance interregional green collaboration, and champion global green partnerships for shared climate solutions.

1.4 Contribution of the Book

1.4.1 Academic Value

Grounded in the strategic imperatives of China's pursuit of carbon neutrality, this book develops a GF–technological empowerment–green development triadic analytical framework and, combining macro-level (city) and micro-level (enterprise) perspectives, systematically examines the mechanisms through which AI, GF, and DI shape the low-carbon transition. Its academic contributions are as follows:

1. Enriching and advancing theoretical research on green and low-carbon transition. This book innovatively incorporates GF and technological empowerment into a unified analytical framework, integrating GF with emerging technologies such as AI and DI. This broadens the theoretical boundaries of sustainable development research and proposes a new analytical paradigm for understanding the synergistic

effects among the financial system, technological change, and environmental governance.
2. Introducing cross-disciplinary perspectives and multi-scale analytical methods into green transformation studies. By integrating theories and methods from economics, management, environmental science, and data science, and applying a dual-scale city–enterprise analysis, the book reveals the transmission mechanisms and differentiated pathways linking macro policy to micro behavior, addressing gaps in existing research regarding cross-scale identification oflow-carbon transition mechanisms.
3. Expanding empirical methods and data applications in green transformation research. Drawing on large-sample data from over 280 cities and 1200 enterprises, the book employs cutting-edge econometric methods such as spatial econometric models, difference-in-differences (DID), and propensity score matching DID (PSM-DID) to empirically evaluate the emission-reduction effects and spatial spillover effects of GF and technological empowerment. These approaches provide replicable empirical strategies and data integration approaches for future studies.

1.4.2 Practical Significance

By focusing on climate change mitigation and low-carbon transition, and aligning closely with China's "30·60" dual carbon strategy, this book conducts an in-depth analysis of the mechanisms through which AI, GF, and DI drive emission reductions, and proposes multilevel policy recommendations. Its practical significance lies in the following:

1. Supporting the implementation of China's dual carbon strategy. The findings provide empirical evidence and mechanism-based insights for achieving China's 2030 carbon peaking and 2060 carbon neutrality goals, identifying feasible pathways for emission reduction driven by technological innovation and financial resource allocation, and offering scientific support for policymaking.
2. Enhancing low-carbon governance capacity of cities and enterprises. At the urban level, the book offers theoretical and practical references for optimizing urban planning, improving green financial instruments, and accelerating DI construction. At the enterprise level, it proposes measures such as AI-driven technological upgrading and GF-guided green

innovation to help firms balance economic performance with sustainable development in the low-carbon transition.

3. Providing transferable experience for developing countries. Based on empirical evidence from China's low-carbon transition, the book distills a "finance–technology–green" synergistic development model that can be adapted to different development stages and institutional contexts, offering valuable lessons for other emerging economies in formulating climate policies and promoting the integration of digital and green development.

References

Alshawi, A. (2022). Artificial intelligence for sustainability: Opportunities and challenges. In *Artificial intelligence for renewable energy and climate change* (pp. 1–31). Wiley. https://doi.org/10.1002/9781119771524.ch1

Broadstock, D. C., Chan, K., Cheng, L. T. W., & Wang, X. W. (2021). The role of ESG performance during times of financial crisis: Evidence from COVID-19 in China. *Finance Research Letters, 38*, Article 101716. https://doi.org/10.1016/j.frl.2020.101716

Friede, G., Busch, T., & Bassen, A. (2015). ESG and financial performance: Aggregated evidence from more than 2000 empirical studies. *Journal of Sustainable Finance & Investment, 5*(4), 210–233. https://doi.org/10.1080/20430795.2015.1118917

Gao, D., Tan, L. F., & Chen, Y. (2025). Smarter is greener: Can intelligent manufacturing improve enterprises' ESG performance? *Humanities & Social Sciences Communications, 12*(1), Article 529. https://doi.org/10.1057/s41599-025-04853-5

Hu, M., Sima, Z., Chen, S., & Huang, M. M. (2023). Does green finance promote low-carbon economic transition? *Journal of Cleaner Production, 427*, Article 139231. https://doi.org/10.1016/j.jclepro.2023.139231

Jiang, H. L., Jiang, P. C., Wang, D., & Wu, J. H. (2021). Can smart city construction facilitate green total factor productivity? A quasi-natural experiment based on China's pilot smart city. *Sustainable Cities and Society, 69*, Article 102809. https://doi.org/10.1016/j.scs.2021.102809

Lee, C. C., & Wang, T. H. (2024). The impact of renewable energy policies on the energy transition: An empirical analysis of Chinese cities. *Energy Economics, 138*, Article 107838. https://doi.org/10.1016/j.eneco.2024.107838

Li, J. C., Lian, G. H., & Xu, A. T. (2023). How do ESG affect the spillover of green innovation among peer firms? Mechanism discussion and performance study. *Journal of Business Research, 158*, Article 113648. https://doi.org/10.1016/j.jbusres.2023.113648

Lou, Y. (2025). Green finance market and carbon neutralization capacity in China. *Humanities and Social Sciences Communications, 12*(1), 1–8. https://doi.org/10.1057/s41599-025-04429-3

Rolnick, D., Donti, P. L., Kaack, L. H., Kochanski, K., Lacoste, A., Sankaran, K., Ross, A. S., Milojevic-Dupont, N., Jaques, N., Waldman-Brown, A., Luccioni, A. S., Maharaj, T., Sherwin, E. D., Mukkavilli, S. K., Kording, K. P., Gomes, C. P., Ng, A. Y., Hassabis, D., Platt, J. C., Creutzig, F., Chayes, J., & Bengio, Y. (2022). Tackling climate change with machine learning. *ACM Computing Surveys (CSUR), 55*(2), 1–96. https://doi.org/10.1145/3485128

Rosenow, J., Fawcett, T., Eyre, N., & Oikonomou, V. (2017). Energy efficiency and the policy mix. *Building Research & Information, 45*(5), 563–575. https://doi.org/10.1080/09613218.2016.1138803

Song, F. (2025). The effects of digital transformation on corporate energy efficiency: a supply chain spillover perspective. *Frontiers in Sustainability, 6*, Article 1567413. https://doi.org/10.3389/frsus.2025.1567413

Tan, Y. F., & Zhu, Z. H. (2022). The effect of ESG rating events on corporate green innovation in China: The mediating role of financial constraints and managers' environmental awareness. *Technology in Society, 68*, Article 101906. https://doi.org/10.1016/j.techsoc.2022.101906

Tang, K. J., & Yang, G. Y. (2023). Does digital infrastructure cut carbon emissions in Chinese cities? *Sustainable Production and Consumption, 35*, 431–443. https://doi.org/10.1016/j.spc.2022.11.022

Wang, Q., & Su, M. (2020). Drivers of decoupling economic growth from carbon emission–an empirical analysis of 192 countries using decoupling model and decomposition method. *Environmental Impact Assessment Review, 81*, Article 106356. https://doi.org/10.1016/j.eiar.2019.106356

Wang, H. L., Jiao, S. P., Ge, C., & Sun, G. L. (2024). Corporate ESG rating divergence and excess stock returns. *Energy Economics, 129*, Article 107276. https://doi.org/10.1016/j.eneco.2023.107276

Zeng, M., & Zhang, W. (2024). Green finance: The catalyst for artificial intelligence and energy efficiency in Chinese urban sustainable development. *Energy Economics, 139*, Article 107883. https://doi.org/10.1016/j.eneco.2024.107883

Zhang, W., & Zeng, M. (2024). Is artificial intelligence a curse or a blessing for enterprise energy intensity? Evidence from China. *Energy Economics, 134*, Article 107561. https://doi.org/10.1016/j.eneco.2024.107561

Zhang, D., Zhang, Z., & Managi, S. (2019). A bibliometric analysis on green finance: Current status, development, and future directions. *Finance Research Letters, 29*, 425–430. https://doi.org/10.1016/j.frl.2019.02.003

Zhang, W., Fan, H., & Zhao, Q. (2023a). Seeing green: How does digital infrastructure affect carbon emission intensity? *Energy Economics, 127*, Article 107085. https://doi.org/10.1016/j.eneco.2023.107085

Zhang, W., Luo, Q., Zhang, Y., & Yu, A. (2023b). Does green credit policy matter for corporate exploratory innovation? Evidence from Chinese enterprises. *Economic Analysis and Policy, 80*, 820–834. https://doi.org/10.1016/j.eap.2023.09.024

Zhang, W., Fan, H., & Zeng, M. (2025). Unleashing the power of artificial intelligence: A game changer for urban energy efficiency in China. *Sustainable Cities and Society, 126*, Article 106372. https://doi.org/10.1016/j.scs.2025.106372

PART I

Urban-Level Analysis

CHAPTER 2

Energy Efficiency and AI Empowerment via Green Finance

2.1 Introduction

With the advancement of urbanization, urban sustainable development has increasingly emerged as a focus in academic discourse (Lee & Lee, 2022; Lee et al., 2023c; Wei et al., 2023; Zeng & Zhang, 2024). As urban centers accommodate growing populations, industries, and transportation networks, they face significant challenges related to energy consumption, pollution emissions, and ecological degradation (Lee & Hussain, 2023; Zeng & Zhang, 2024). Over the past four decades, China has experienced extensive urbanization characterized by high investment and low efficiency. The prevailing approach to urban development, which relies heavily on resource investment, is unsustainable (Wan et al., 2024; Zeng & Zhang, 2024). Consequently, Chinese cities are confronting severe urban challenges and are under growing pressure to achieve sustainable development. Thus, transforming urban development models and realizing sustainable urban development have become urgent issues that must be addressed in China. The existing research on promoting urban sustainable development has been extensively and comprehensively discussed. The

The core content of this chapter is derived from this article: Zeng, M., & Zhang, W. K. (2024). Green finance: The catalyst for artificial intelligence and energy efficiency in Chinese urban sustainable development. *Energy Economics, 139*, Article 107883. https://doi.org/10.1016/j.eneco.2024.107883

W. Zhang et al., *Green Transformation in the Digital Age*,
https://doi.org/10.1007/978-981-95-9121-3_2

factors affecting urban sustainability that the existing research focus on involve digital economy (Liu et al., 2024), urban governance (Shu et al., 2023), urban expansion (Wei et al., 2023), infrastructure construction (Zhang et al., 2023a), and so on.

Among them, artificial intelligence (AI) plays a crucial role in the advancement of urban sustainable development (Wang et al., 2023; You et al., 2025; Zeng & Zhang, 2024; Zhang et al., 2025). AI is a machine system that provides decision-making recommendations based on specific human instructions, by simulating human behavior and thought patterns (Babina et al., 2024). AI has the potential to drive advancements in production technology, optimize the layout of industrial structures, and enhance energy efficiency, ultimately leading to urban sustainable development (Luo & Feng, 2024). Furthermore, energy efficiency (EE) is often seen as an explicit indicator to monitor urban sustainable development (Song et al., 2024). EE refers to the ratio between the target energy input and the actual energy input, which is used to mark the level of energy utilization in a region (Hu & Wang, 2006). Energy is the foundation of urban development, and its utilization efficiency directly reflects the extent of urban sustainable development (Wu et al., 2023).

In recent years, green finance (GF) has garnered significant attention for its potential to advance urban sustainable development. GF refers to financial services such as bonds, investments, and governance initiatives undertaken by governments, banks, and enterprises to support economic activities in sectors including green energy, transportation, and construction. The ultimate goals of GF are to promote environmental protection, resource conservation, and sustainable development (Lv et al., 2021; Zeng & Zhang, 2024). Consequently, countries around the world are competing to implement stimulus policies, strengthen institutional frameworks, and refine financial systems to promote GF (Su et al., 2022). In China, GF is gaining traction, as evidenced by the country's cumulative issuance of green bonds, which ranks second globally (Lee et al., 2023c). As illustrated in Fig. 2.1,[1] China issued a total of 802 green bonds with a cumulative issuance amount of 1118.05 billion RMB in 2023. This figure reflects stability compared to the previous year and marks a significant increase relative to 2021.

[1]The data is from Economic Information Daily, whose website is http://www.jjckb.cn/2024-01/22/c_1310761333.htm

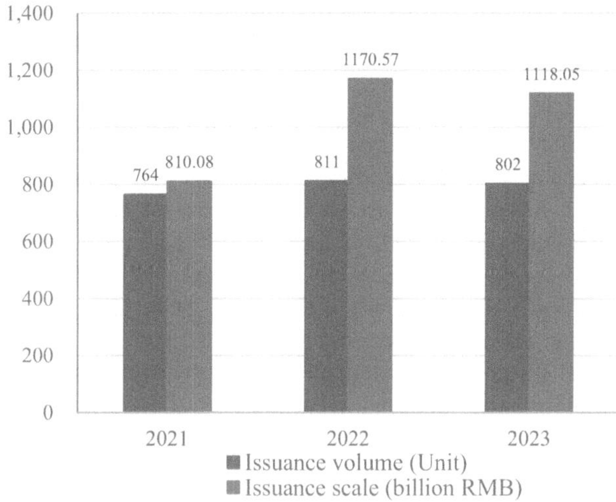

Fig. 2.1 Current status of green bonds in China

Focusing on urban sustainable development, existing studies have extensively explored the impact of GF. Recent studies have generated growing interest in the effects of GF on environmental performance. A body of research has examined the influence of GF in this context. Such studies have shown that GF can effectively allocate resources to environmentally friendly economic activities, leading to a reduction in carbon emissions, the mitigation of environmental degradation, and ultimately an improvement in overall environmental quality (Bakry et al., 2023; Afzal et al., 2022; Al-Mamun et al., 2022). Meanwhile, the role of GF in promoting green innovation has also been examined. They argue that GF facilitates high-quality green innovation among enterprises while constraining their strategic innovation behavior (Yu et al., 2021; Huang et al., 2022a). What is more, GF promotes green innovation not only within local contexts but also in neighboring regions (Huang et al., 2022b). In addition, some studies have investigated the GF–EE nexus. Their findings show that GF can enhance EE by fostering green technology innovation and optimizing energy structures (Lee et al., 2023a; Liu et al., 2022; Wei et al., 2025), with its impact surpassing that of fintech and financial inclusion (Guo et al., 2023). Meanwhile, Li et al. (2025) evidence that green financial instruments, including green credit, investment, insurance, and subsidies, can effectively reduce energy intensity and overall energy consumption, whereas the impacts of green equity and bonds are less pronounced. In contrast, Wang and Wang (2022) conclude that GF may increase EE in Chinese eastern regions, but it lacks a similar effect in the central and western regions. Despite the considerable progress in GF-related research, several shortcomings remain. First, most studies have focused on countries or provinces, with limited exploration at the city level regarding the impact of GF. Second, no literature addressing the impact of GF on AI has been identified. Finally, few studies have investigated the influence of GF on EE, particularly its spatial spillover effects.

To address the aforementioned shortcomings, this chapter is designed to explore how GF drives urban sustainable development through AI and EE, using 282 Chinese cities' datasets from 2014 to 2019. The findings indicate that GF contributes to the enhancement of urban AI levels. These conclusions remain robust even when alternative measurements are considered, the IV method is employed, the interference of the GF policies is eliminated, the municipality samples are excluded, and the doubly debiased Lasso (DDL) regression is carried out. Additionally, the variation of these impacts across different city typologies has also been discussed. Furthermore, it explores the GF–EE nexus, revealing that GF not only improves EE but also generates positive spatial spillover effects on EE. Figure 2.2 illustrates the roadmap of this chapter. GF is dedicated to utilizing the financial

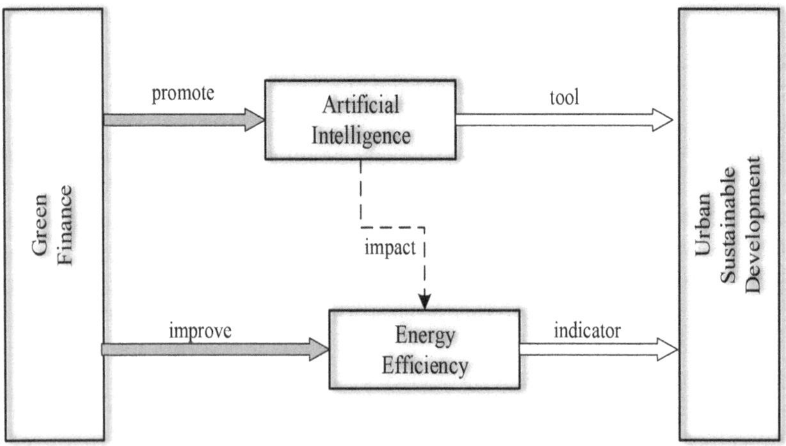

Fig. 2.2 Roadmap for research

system to achieve urban sustainable development, with AI serving as a tool to realize this objective (Wang et al., 2023), while EE represents GF's achievements in promoting urban sustainability (Song et al., 2024).

2.2 Theoretical Analysis

2.2.1 The Effect of Green Finance on AI

Existing studies have primarily examined the role of GF in technological progress (Bai et al., 2025; Lee & Lee, 2022; Lee et al., 2023c; Lei et al., 2025; Zhang et al., 2023b), with limited research addressing its impact on AI (Chen et al., 2024). Numerous investigations have consistently demonstrated the positive influence of GF in driving technological advancement. For instance, Xu et al. (2023) utilized China's green finance pilot policies to reveal that GF facilitates technology exchange and cooperation among enterprises, thereby stimulating technology proliferation and fostering innovation within the industry. Similarly, Cui et al. (2023) found that GF alleviates financial constraints for Chinese enterprises, leading to improved financial support for technological development and subsequently accelerating overall innovation. Additionally, Huang et al. (2022b) observed that GF can stimulate innovation outputs based on provincial samples in China. Given these findings, it is highly likely that GF will contribute to the advancement of urban AI. As a cutting-edge technology, AI epitomizes urban technological progress, suggesting that GF can drive the development of AI.

In fact, GF can significantly facilitate urban AI from both supply and demand perspectives. Figure 2.3 illustrates the theoretical rationale. On the supply side, GF provides essential resources for AI development in urban settings, which often face financing challenges due to their high-risk nature (Zhang & Zeng, 2024). Green financial instruments are vital in channeling capital toward advanced industries and cutting-edge technologies. This substantial investment is crucial for driving innovation and progress in urban AI development (Madaleno et al., 2022). For instance, the "Green Bond Support Project Catalog (2021 edition)"[2] highlights

[2] It was jointly issued by the People's Bank of China, the China Development and Reform Commission, and the China Securities Regulatory Commission. The website can be accessed at: http://www.pbc.gov.cn/goutongjiaoliu/113456/113469/4236341/2021042115215612655.pdf

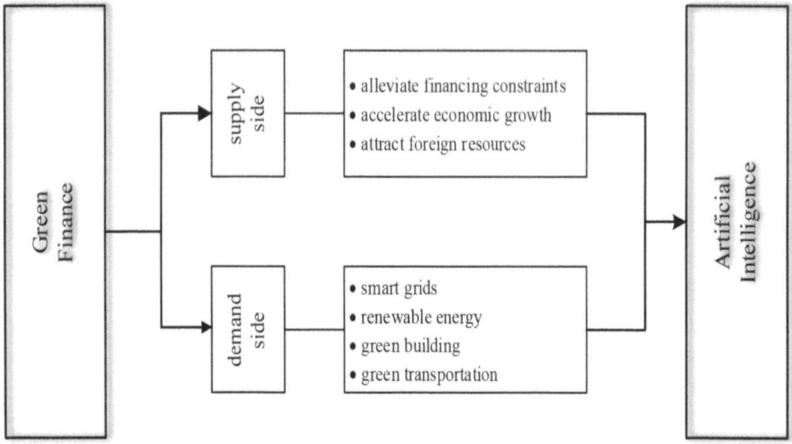

Fig. 2.3 Theoretical mechanisms of the effect of GF on AI

key areas of support, including intelligent systems in power grids, transportation, construction, and other sectors. This illustrates how GF alleviates financing constraints on AI development, thereby significantly enhancing urban AI capabilities. Additionally, GF has made notable contributions to urban economic growth and the overall strength of cities (Lee & Lee, 2022). These initiatives have greatly increased the attractiveness of urban areas for foreign investments, technical talent, and AI-related collaborations. In summary, GF helps advance AI supply.

On the demand side, GF advocates for the advancement of emerging industries, which demonstrate a significant need for AI. Numerous studies indicate that the development of GF has spurred the emergence of a new class of frontier industries and has enabled the enhancement and modernization of industrial frameworks (Zhang, 2023; Wang & Wang, 2021). For instance, Ma et al. (2024) have elucidated the significance of green bonds in bolstering urban financing mechanisms and substantiated their favorable influence on the advancement of urban smart grids. Debrah et al. (2022) argued that GF has addressed the funding gap resulting from inadequate investment in green buildings, thereby facilitating the intelligent transformation of the construction industry toward sustainability. The emerging industries are frequently distinguished by significant technological requirements, where AI like cloud computing and machine learning play a crucial role in their advancement (Zeba et al., 2021). Consequently, the rise of these industries has led to a substantial increase in the demand for AI, creating significant market opportunities. In summary, GF is committed to promoting green and intelligent development across various sectors, including energy, transportation, and construction, which has markedly increased the demand for AI technology.

Therefore, we propose the following hypothesis:

Hypothesis 2.1: GF can significantly enhance urban AI level.

2.2.2 *The Effect of Green Finance on Energy Efficiency*

Figure 2.4 illustrates the theoretical rationale behind how GF affects EE. In principle, GF contributes to the enhancement of EE. Firstly, GF promotes technological advancement, particularly in the field of AI, which provides significant benefits in improving EE. The positive relationship between finance and technological progress has been well established in the academic literature (Yang & Ni, 2022). Technological advancements have optimized production processes and facilitated energy conservation, thereby enhancing EE. Secondly, GF fosters financial innovation, which further increases EE. Through the advancement of financial development,

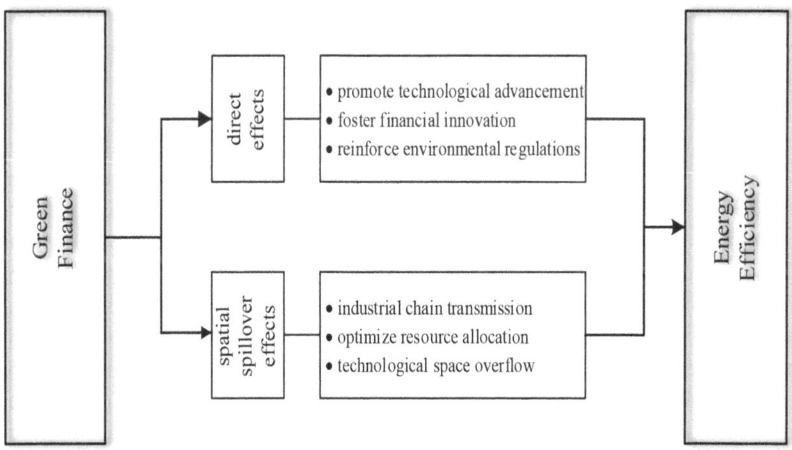

Fig. 2.4 Theoretical mechanisms of the effect of GF on EE

innovation in financial products, and the updating of institutional frameworks, GF ultimately seeks to enhance EE (Wang & Wang, 2022). Thirdly, GF reinforces environmental regulations, thereby enhancing EE. GF is committed to strengthening environmental protection through financial mechanisms, promoting the disclosure of environmental information, and raising standards for environmental access. These initiatives are likely to compel cities to develop more efficient energy systems, thus increasing EE (Song et al., 2021). Consequently, the following hypothesis is developed:

Hypothesis 2.2: GF can significantly enhance EE.

The new economic geography theory posits that economic activities often generate spatial spillover effects, resulting in interactive outcomes among cities (Zhang et al., 2023c). Consequently, local EE is influenced not only by GF within a city but also by the GF in neighboring cities. Economic activities across cities are intricately interconnected, often functioning as upstream or downstream components of the industrial chain (Ma et al., 2023). Thus, improvements in the EE of local cities, driven by their GF, can extend to other cities within the same industrial chain. Neighboring cities can benefit from the GF present in local cities, contributing to an overall enhancement of EE across the region. Moreover, GF embodies characteristics of traditional finance that can dismantle financial barriers between regions, facilitating the unrestricted flow of capital. This dynamic enhances resource allocation efficiency and concurrently improves interregional EE (Lee et al., 2023b; Zhang et al., 2023d). Specifically, GF in local cities can result in a reduction of available capital for neighboring cities (Zhao et al., 2023). Consequently, enterprises in these neighboring cities that exhibit low energy efficiency may be driven out of the market due to insufficient capital, ultimately leading to an improvement in the overall EE of those cities. Furthermore, GF fosters technological advancement, which generates notable spatial spillover effects. The advancement of technology can bring benefits to local cities and extend its advantages to neighboring areas, showcasing evident spillover effects (Huo et al., 2022). Based on this, GF has the potential to drive technological advancement in neighboring cities, thereby elevating their EE. In summary, the positive impact of GF on EE demonstrates a spatial spillover phenomenon, where GF can improve EE in both local and neighboring cities.

Therefore, we propose the following hypothesis:

Hypothesis 2.3: The enhancement effect of GF on EE exhibits the characteristics of spatial spillover.

2.3 Research Design

2.3.1 Data

By leveraging the 2014–2019 Chinese city data, we explore how GF affects AI. The AI-related data is from the International Federation of Robotics (IFR).[3] It provides industry-level data rather than city-level data, so this chapter needs to decompose the data to the city level. Since IFR only published industrial robot data up to 2019, the end of the sample period is determined to be 2019. The GF-related data is from the Wind Economic Database (Wind EDB).[4] Prior to 2014, the issuance scale of green bonds, which measure GF, was very limited. Hence, the sample period spans from 2014 to 2019. Besides, the other data is from the Chinese Research Data Services Platform (CNRDS).[5]

2.3.2 Variable

The Explained Variable Most of the existing studies have utilized exposure to industrial robots as a measure of AI. For instance, Liu et al. (2021) employed industrial robot exposure to gauge AI and conducted an analysis of its impact on industry energy intensity. Similarly, Li et al. (2023) used enterprise industrial robot exposure as a metric for AI and investigated how AI affects enterprise EE with this measure. Besides, Yin et al. (2023) utilized the provincial industrial robot data in China to quantify AI and assess its impact on enhancing green innovation. In addition to this approach, Li et al. (2024) utilized the number of AI enterprises within the city as a metric for assessing urban AI level.

Industrial robots are defined as automated devices used in industrial applications that can be programmed to replicate human intelligence and possess the characteristics of autonomy, adaptability, and automation, serving as a primary form of AI utilized in production processes (Acemoglu & Restrepo, 2020). And exposure to industrial robots can serve as a

[3] IFR: https://ifr.org/

[4] Wind: https://www.wind.com.cn/

[5] CNRDS: https://www.cnrds.com/

reliable indicator for the advancement of AI (Zhang & Zeng, 2024). Hence, this chapter adopts exposure to industrial robots as a proxy metric for measuring AI. Exposure to industrial robots is typically quantified by the ratio of industrial robots to employment levels. Drawing from the work of Zeng and Zhang (2024) and Zhang and Zeng (2024), this chapter calculates urban exposure to industrial robots by decomposing the exposure at the industry level. Hence, AI is obtained by Eq. 2.1:

$$AI_{it} = \sum_{j=1}^{J} \frac{\text{empl}_{i,j,t=2005}}{\text{empl}_{i,t=2005}} \times \frac{\text{robot}_{jt}}{\text{empl}_{j,t=2005}} \tag{2.1}$$

where i, j, and t indicate city, industry, and year, respectively; robot is the number of industrial robots and empl is the employment.

The formulation of Eq. 2.1 is in line with the Bartik instrumental variable method, allowing for precise estimation of urban exposure to industrial robots using industry-level data. This chapter calculates urban exposure to industrial robots based on both newly installed robots and the stock of existing robots, denoting them as AI_1 and AI_2, respectively.

The Explanatory Variable GF encompasses a broad spectrum of activities related to environmental sustainability (Lv et al., 2021). The academic community typically adopts a weighted approach to develop an index system that evaluates GF across multiple green dimensions, including bonds, credit, insurance, and investments. However, this methodological framework is primarily applicable at the provincial level and poses challenges for city-level analysis due to the scarcity of relevant data (Yue et al., 2024). City-level studies often rely on data from a singular dimension, with green bonds being the most frequently employed metric due to their distinct advantages. Firstly, green bonds encapsulate the core essence of green finance and represent a quintessential form of this financial model (Al-Mamun et al., 2022). These securities are issued by banks or enterprises following stringent procedures and compliance with policy requirements, specifically to finance green projects, green industries, and other environmentally sustainable economic activities (Naeem et al., 2021). Secondly, the comprehensive disclosure of data related to green bonds effectively addresses the research needs at the city level. Consequently, following Zeng and Zhang (2024) and Zhu et al. (2021), this chapter employs the volume of green bonds issued in each city as the principal

Table 2.1 Descriptive statistics

Variables	*Obs*	*Mean*	*S.D.*	*Min*	*Max*
	(1)	*(2)*	*(3)*	*(4)*	*(5)*
Core variables					
AI_1 (units/thousand workers)	1692	0.875	0.481	0.114	2.853
AI_2 (units/thousand workers)	1692	3.550	2.267	0.435	14.237
GF (green bond issuance scale, 100 million RMB, in log)	1692	0.320	1.025	0	6.865
Control variables					
Growth rate of urban GDP (%)	1692	0.547	1.031	−6.717	5.590
Proportion of the secondary industry in outputs	1692	0.446	0.101	0.117	0.755
Ratio of fiscal revenue to GDP	1692	0.079	0.025	0.023	0.202
Government spending on education as a share of GDP	1692	0.035	0.018	0.009	0.148
Government spending on technology as a share of GDP	1692	0.003	0.003	0	0.063

indicator for evaluating urban GF. Additionally, this chapter employs the green bond issuance per unit of GDP to conduct a robustness test.

The Control Variables Several city-level variables are introduced in this chapter to control potential confounding effects on the GF–AI nexus, as presented in Table 2.1, which displays their basic statistical information. In Column (2), the average density of installed industrial robots in Chinese cities from 2014 to 2019 is 0.875 per 1000 workers, with a stock of 3.550 per 1000 workers. These measurements align with Wei et al. (2020), who reported an average installation rate of 0.407 units per 1000 urban workers in China during 2012, 2014, and 2016.

2.3.3 Method

To examine the effect of GF on AI, Eq. 2.2 is adopted:

$$AI_{it} = \alpha_0 + \beta GF_{it} + \eta C_{it} + u_i + v_t + \varepsilon_{it} \tag{2.2}$$

where C represents the controls; u_i and v_t represent the city and year fixed effects (FE) respectively; ε_{it} is the random interference term.

2.4 Findings

2.4.1 *Baseline Results*

Table 2.2 displays the outcomes of Eq. 2.2. All coefficients associated with GF are statistically significant. That is, GF can notably improve Chinese urban AI, thus supporting Hypothesis 2.1. Particularly noteworthy is the coefficient (0.020) in the second column, which means a one-unit increase in GF results in a 0.020 increase in AI_1. That is, an increase of one standard deviation in GF leads to an increase of 0.043 (0.020 × 1.025/0.481) standard deviations in the installation of industrial robots per thousand workers in Chinese cities. Similarly, 0.138 for AI_2 indicates that every additional standard deviation of GF can increase the stock of industrial robots per thousand workers in Chinese cities by 0.062 (0.138 × 1.025/2.267) standard deviations.

These findings align with the work of Lee et al. (2023c), which demonstrates that GF facilitates green innovation through the implementation of green bond policies. Specifically, GF can enhance urban AI development from both supply and demand perspectives. As an emerging financial instrument designed to promote low-carbon and efficient economic growth, GF has the potential to effectively mobilize resources for the advancement of AI, thereby addressing the persistent issue of inadequate funding that has impeded urban AI development (Zhang & Zeng, 2024). Furthermore, GF facilitates the development of nascent industries, leading to the growth of high-technology sectors and the enhancement of industrial frameworks (Wang & Wang, 2021). These emerging industries

Table 2.2 Baseline regression results

	AI_1		AI_2	
	(1)	*(2)*	*(3)*	*(4)*
GF	0.022***	0.020***	0.156***	0.138***
	(0.005)	(0.005)	(0.035)	(0.033)
Controls		√		√
City FE	√	√	√	√
Year FE	√	√	√	√
Observations	1692	1692	1692	1692
R^2	0.952	0.954	0.918	0.921

***$p < 0.01$. The values in parentheses are robust standard errors, clustered at the city level

possess significant technological demands, with AI technologies such as cloud computing and big data playing a pivotal role, thereby creating substantial market opportunities for urban AI.

2.4.2 Robustness Tests

Alternative Measure for AI In the context of assessing AI, this chapter presents a new metric called "AI enterprise density," which quantifies the concentration of AI enterprises per one thousand urban residents. Data on AI enterprises were collected through a comprehensive search of the Tianyancha website, focusing on enterprise names and business scopes to identify connections with AI-related keywords. The findings serve to confirm earlier conclusions, as the positive correlation between GF and AI enterprise density in first column of Table 2.3 reveals GF promotes AI development in China.

Alternative Measure for GF Previously, GF was measured based solely on the absolute number of green bond issuances, without considering the

Table 2.3 Results of robustness tests (Part A)

	Alternative measure for AI	*Alternative measure for GF*		*IV method*	
	AI enterprise density	AI_1	AI_2	AI_1	AI_2
	(1)	*(2)*	*(3)*	*(4)*	*(5)*
GF	0.031***				
	(0.007)				
GF per unit of GDP		0.017***	0.124***		
		(0.004)	(0.030)		
GF_IV				0.076***	0.538***
				(0.022)	(0.149)
Controls	√	√	√	√	√
City FE	√	√	√	√	√
Year FE	√	√	√	√	√
Observations	1692	1692	1692	1662	1662
R^2	0.915	0.954	0.921	0.954	0.922

***$p < 0.01$. The values in parentheses are robust standard errors, clustered at the city level. The sample size in Column (4) and (5) is smaller than in the others because the sample of provinces with only one city has been dropped

economic development status of the city. In this chapter, however, GF is assessed from a relative perspective, specifically by taking the logarithm of the scale of green bond issuance per unit of city GDP. Column (2) of Table 2.3 illustrates the effect of GF per unit of GDP on AI_1, while Column (3) displays its effect on AI_2. The coefficients in these columns align with the aforementioned findings, further demonstrating that GF enhances AI in Chinese cities.

IV Method Theoretically, AI has the potential to enhance the level of GF in Chinese cities. The advancements in financial technology driven by AI are expected to accelerate the development of GF within urban areas. If this hypothesis is validated, it may introduce issues of reverse causality and bias the regression results when examining the relationship between GF and AI. To address this, we construct an instrumental variable for GF (GF_IV) by utilizing its spatial lag term. Specifically, GF_IV for a given city is defined as the average GF of other cities within the same province. The last columns of Table 2.3 are the results for the reduced-form model using GF_IV. The findings indicate that GF_IV significantly increases both AI_1 and AI_2, thereby reaffirming the conclusions drawn earlier.

Eliminating the Interference of the GF Policies In 2017, the Chinese government declared the creation of pilot zones aimed at implementing reforms and fostering innovation in green finance, henceforth denoted as GF policies. Eight pilot cities were initially selected: Quzhou, Huzhou, Guangzhou, Hami, Changji, Karamay, Guian New Area, and Ganjiang New Area. In 2019, Lanzhou was added as an additional pilot city. These GF policies are expected to promote green finance initiatives in the designated cities. However, the existing studies do not account for the potential influence of these GF policies. To address this issue, the samples from these pilot cities were excluded, and a regression analysis based on Eq. 2.2 was conducted again. As shown in the first two columns of Table 2.4, the results reconfirm the above conclusions.

Excluding the four Municipality Samples The four municipalities of Beijing, Tianjin, Shanghai, and Chongqing receive greater policy and resource support compared to other cities, facilitating advancements in AI through GF. Consequently, the AI capabilities of these four municipalities are significantly higher than those of other cities, which may complicate efforts for GF to enhance AI in less-developed areas. This discrepancy

Table 2.4 Results of robustness tests (Part B)

	Eliminating the interference of the GF policies		*Excluding the four municipality samples*		*DDL regression*	
	AI_1	AI_2	AI_1	AI_2	AI_1	AI_2
	(1)	*(2)*	*(3)*	*(4)*	*(5)*	*(6)*
GF	0.021***	0.142***	0.022***	0.155***	0.023***	0.159***
	(0.005)	(0.037)	(0.005)	(0.035)	(0.005)	(0.025)
Controls	√	√	√	√		
City FE	√	√	√	√	√	√
Year FE	√	√	√	√	√	√
Observations	1644	1644	1668	1668	1662	1662
R^2	0.953	0.920	0.954	0.921	0.954	0.922

***$p < 0.01$. The values in parentheses are robust standard errors, clustered at the city level. The regression results for DDL are highly sensitive to the control variables. Therefore, in order to avoid any interference from the control variables, they are not included in the DDL regression analysis

raises the possibility that the aforementioned conclusion may inaccurately estimate the impact of GF on AI. To address this concern, this chapter excludes these municipalities and reevaluates the relationship between GF and AI using samples that do not include them. Columns (3) and (4) of Table 2.4 demonstrate a consistently positive relationship, confirming the robustness of our conclusions.

DDL Regression To mitigate potential confounding effects arising from the omission of uncontrollable variables, which could undermine the validity of the conclusions regarding the GF-AI relationship, this chapter employs the DDL method for analysis. This approach reduces estimation biases caused by confounding factors by introducing a specific matrix (Guo et al., 2022). The last columns of Table 2.4 present the results. The coefficients for GF are slightly higher than those in the baseline regression presented in Table 2.2 (0.023 vs. 0.022; 0.159 vs. 0.156). Hence, there is no significant bias due to omitted variables, thereby reinforcing the credibility of these conclusions.

Table 2.5 Results of heterogeneity analysis

	Region heterogeneity		*Resource heterogeneity*		*Scale heterogeneity*		*AI-level heterogeneity*	
	Western	*Eastern and Central*	*Resource-dependent*	*Non–resource-dependent*	*Type I: pop ≥ 5 million*	*Type II: pop < 5 million)*	*High-AI-level*	*Low-AI-level*
	(1)	(2)	(3)	(4)	(5)	(6)	(7)	(8)
Part A: The dependent variable is AI_1.								
GF	0.009	0.021***	0.013	0.011**	0.013**	0.025***	0.002	0.017***
	(0.006)	(0.005)	(0.011)	(0.005)	(0.005)	(0.008)	(0.004)	(0.004)
R^2	0.958	0.954	0.951	0.960	0.965	0.949	0.973	0.973
Part B: The dependent variable is AI_2.								
GF	0.056	0.154***	0.126	0.083**	0.082**	0.215***	0.028	0.110***
	(0.034)	(0.040)	(0.089)	(0.035)	(0.035)	(0.066)	(0.031)	(0.019)
R^2	0.923	0.925	0.918	0.930	0.939	0.914	0.958	0.961
Controls	√	√	√	√	√	√	√	√
City FE	√	√	√	√	√	√	√	√
Year FE	√	√	√	√	√	√	√	√
N	510	1182	684	1008	606	1086	846	846

***$p < 0.01$. **$p < 0.05$. The values in parentheses are robust standard errors, clustered at the city level. According to the Notice of the State Council on Adjusting the Criteria for the Division of City Scale released in 2014, this chapter gets two categories based on the number of permanent urban residents: Type I cities (population ≥5 million) and Type II cities (population < 5 million). Since the coefficients in Columns (5) and (6) are significantly positive, this chapter further adopts the Seemingly Unrelated Test to identify the significance of the numerical differences between them. The test results showed that P values for AI_1, namely 0.012 vs. 0.029, and AI_2, namely 0.081 vs. 0.242, are 0.091 and 0.007, respectively. These results evidence that the coefficients in Column (6) are significantly greater that those in Column (5)

2.4.3 Heterogeneity Analysis

We analyzed heterogeneity based on region, resources, scale, and AI level. Their results are presented in Table 2.5.

Region Heterogeneity Economic conditions and technological advancements in China exhibit significant disparities, with the eastern and central regions demonstrating notable advantages over their western counterparts. These more developed regions attract greater capital investment, technological innovations, and skilled individuals. As a result, the impact of GF on AI may differ between the western and central cities and the eastern cities.

Column (1) shows positive but nonsignificant coefficients, while those in Column (2) are significantly positive. Hence, GF can benefit AI advancement in the eastern and central cities, whereas its effect on the western cities is less apparent. In fact, the eastern and central cities possess superior foundational conditions, including advanced equipment, more technical talent, and policy support. These factors play a catalytic role in GF's promotion of AI development and strengthen the enhancement effect of AI by GF. Conversely, the limited availability of AI technical expertise, insufficient technological infrastructure, and constrained market in the western cities have significantly impeded the effect of GF. Consequently, GF has not substantially influenced the development of AI in western cities.

Resource Heterogeneity Significant differences exist between resource-dependent cities and those that are not, particularly regarding their resource stock, industrial composition, and growth trajectories. These differences may influence the link between GF and AI. Resource-dependent cities are identified according to the National Sustainable Development Plan for Resource-Dependent Cities issued by the Central People's Government of China.[6] Column (3) is for resource-dependent cities, while Column (4) is for non–resource-dependent cities.

In Column (3), the coefficients of GF are positive but not statistically significant for resource-dependent cities, indicating that GF does not substantially contribute to enhancing AI levels in these areas. Conversely, non–resource-dependent cities exhibit significant positive coefficients, suggesting that GF has a meaningful impact on AI development in these cities.

[6] The URL is https://www.gov.cn/zwgk/2013-12/03/content_2540070.htm

Hence, the impact of GF on AI is predominantly noticeable in cities that are not reliant on available resources. Resource-dependent cities tend to exhibit a strong reliance on natural resources, characterized by high energy consumption and extensive development. These cities often experience the "resource curse," which results in a developmental path that is heavily reliant on resource extraction, making it difficult to achieve advancements in cutting-edge technology. As a result, GF has not played a crucial role in significantly enhancing AI capabilities in resource-dependent cities.

Scale Heterogeneity The size of cities plays a crucial role in determining the extent to which the scale effect of their economies can be actualized, and the influence of GF on AI may exhibit scale heterogeneity. In accordance with the Notice of the State Council on Adjusting the Criteria for the Division of City Scale released in 2014,[7] this chapter categorizes cities into two types based on the number of permanent urban residents: Type I cities (population ≥5 million) and Type II cities (population <5 million).

Columns (5) and (6) both show significant positive coefficients, indicating that GF enhances AI levels in both Type I and Type II cities. However, the coefficients for the latter are much larger than the former ones. And this intergroup coefficient difference is confirmed by the Seemingly Unrelated Test. These findings suggest the promotion of AI is more pronounced in Type II cities compared to Type I cities. This disparity may be attributed to the crowding effect observed in Type I cities. As economic and cultural hubs, Type I cities typically have access to advanced technology, substantial capital, and abundant resources, which foster their development but simultaneously contribute to a crowding effect. After decades of growth, the "urban disease" faced by China's Type I cities has become increasingly pronounced (Zhang et al., 2023c). Consequently, the crowding effect in Type I cities limits the potential of GF to promote AI.

AI-Level Heterogeneity The impact of GF on AI may be contingent upon the existing level of urban AI. To investigate this, the sample was evenly divided into two groups based on the average AI level over the years: high-AI-level cities and low-AI-level cities. The influence of GF on AI was subsequently analyzed for each group separately. As indicated in Column (7), the coefficients of GF for high-AI-level cities are positive but not

[7] The URL is https://www.gov.cn/zhengce/content/2014-11/20/content_9225.htm

statistically significant, suggesting that GF has not substantially enhanced their AI capabilities. Conversely, Column (8) reveals that the coefficients of GF for low-AI-level cities are significantly positive, indicating a notable impact of GF on their AI development. The observed disparity in the effects of GF on AI advancement between high-AI-level and low-AI-level cities may arise from the greater challenges associated with improving AI levels in the former, in contrast to the relatively more straightforward task faced by the latter due to their larger demand space.

2.4.4 Further Analysis on the Effect of Green Finance on Energy Efficiency

Previously, we conducted an analysis of the impact of GF on urban AI levels. Here, this chapter further examines how GF impacts EE, with a specific focus on its spatial spillover effect.

The Measurement of EE There are two measurement methodologies for EE as discussed by Hong et al. (2022). The first methodology is the single-factor approach, which quantifies EE by economic consequences of energy inputs (Wang et al., 2022). This approach focuses exclusively on energy and economic outcomes, thereby neglecting other inputs that contribute to production activities and disregarding pollutant emissions (Qu et al., 2020). In contrast, the total-factor approach evaluates EE by incorporating multiple inputs and outputs, rendering it a more scientifically rigorous methodology (Sun et al., 2019; Wang & Shao, 2023). Consequently, this chapter adopts the total-factor approach to measure EE.

Building on the framework established by Chen and Wang (2022), this chapter investigates three primary input factors: urban aggregate energy consumption, urban fixed capital stock, and the urban employed labor force. On the output side, the analysis focuses on urban real GDP as a desired output, while urban emissions of carbon dioxide, industrial sulfur dioxide, soot, and effluent are considered undesired outputs. To assess Chinese EE, this chapter employs the super-efficiency Data Envelopment Analysis (DEA) method, as recommended by Liao and He (2018). This method provides several advantages, including the ability to accommodate variables of varying dimensions without requiring weight assumptions.

Furthermore, it facilitates a more accurate identification of efficient units and reduces the issue of slack bias induced by traditional DEA (Zhou et al., 2012). The detailed calculation process is outlined as follows:

$$\min\left[\theta - \varepsilon\left(\sum_{i=1}^{m} s_i^- + \sum_{r=1}^{s} s_r^+\right)\right]$$

$$s.t.\begin{cases} \sum\limits_{j=1, j\neq j_0}^{n} X_{ij}\lambda_j + s_i^- = \theta X_{ij_0}, i = 1,2,\ldots,m \\ \sum\limits_{j=1, j\neq j_0}^{n} Y_{rj}\lambda_j - s_r^+ = Y_{rj_0}, r = 1,2,\ldots,s \\ \lambda_j, s_i^-, s_r^+ \geq 0, j = 1,2,\ldots,n \end{cases} \tag{2.3}$$

where θ captures EE of city j_0; X and Y are input and output vectors; λ_j defines the weight coefficients; and s_i^- and s_r^+ are slack variables.

Spatial Autocorrelation Analysis The presence of spatial autocorrelation among variables is a prerequisite for performing spatial analysis (Fingleton & Gallo, 2008). This phenomenon indicates that variables in local cities are correlated with those in neighboring cities. Before conducting spatial autocorrelation tests, spatial weight matrices need to be introduced, which represent the complex spatial relationships among cities. These matrices are crucial for quantifying the proximity and interconnectivity between cities, thereby effectively defining the concept of "neighboring." In accordance with Zhang et al. (2023c), this chapter develops three spatial weight matrices: the contiguity matrix (denoted as $\mathbf{W}_1$), the economic matrix (denoted as $\mathbf{W}_2$), and the nested matrix (denoted as $\mathbf{W}_3$), as specified in Eq. 2.4.

$$\mathbf{W}_1 : w_{ij} = \begin{cases} 1, i \text{ is adjacent to } j, \text{and } i \neq j \\ 0, \text{else} \end{cases}$$

$$\mathbf{W}_2 : w_{ij} = \begin{cases} \dfrac{1}{\left|\mathrm{GDP}_i - \mathrm{GDP}_j - 1\right|}, i \neq j \\ 0, i = j \end{cases}$$

$$\mathbf{W}_3 : w_{ij} = \begin{cases} \dfrac{1}{\left|\mathrm{GDP}_i - \mathrm{GDP}_j - 1\right| \times e^{d_{ij}}}, i \neq j \\ 0, i = j \end{cases} \tag{2.4}$$

Table 2.6 Global Moran's Index values of EE

	W_1	W_2	W_3
	(1)	*(2)*	*(3)*
2014	0.018	0.054*	0.077**
2015	0.070*	0.041	0.074**
2016	0.070*	0.049*	0.067**
2017	0.172***	0.104***	0.141***
2018	0.190***	0.139***	0.172***
2019	0.214***	0.098***	0.139***

***$p < 0.01$. **$p < 0.05$. *$p < 0.1$

where GDP is the average urban 2006–2019 GDP; d is the geographical distance between cities.

Table 2.6 presents the results of the Global Moran's index for EE. As shown, these values exhibit positive signs, with most of them demonstrating statistical significance. Hence, there is spatial autocorrelation in EE, and the implementation of spatial analysis is both rational and indispensable in this chapter.

Figure 2.5 displays the scatterplots of the local Moran's Index for the year 2019. As shown, most points fall in the first or third quadrant. These results demonstrate a clear pattern of high-value clusters coexisting with other high-value cities, and low-value clusters coexisting with other low-value cities. In conclusion, these findings strongly underscore the importance of conducting spatial analysis to evaluate the influence of GF on EE.

Econometric Models This chapter uses Tobit model to verify the direct effects of GF on EE:

$$\mathrm{EE}_{it}^{\ *} = \alpha_0 + \beta \mathrm{GF}_{it} + \eta C_{it} + u_i + v_t + \varepsilon_{it} \tag{2.5}$$

where EE^* represents the potential value of EE; β captures the direct effects; the other symbols carry identical connotations as those in Eq. 2.2.

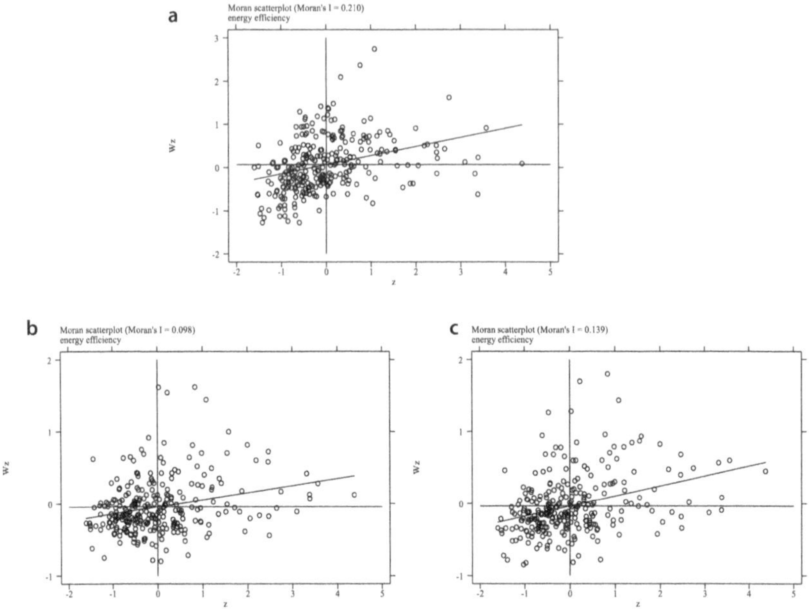

Fig. 2.5 Local Moran's Index scatterplot of UEE in 2019. (**a**) The scatterplot based on $\mathbf{W}_1$. (**b**) The scatterplot based on $\mathbf{W}_2$. (**c**) The scatterplot based on $\mathbf{W}_3$

Its spatial spillover effects are tested by these spatial models, including the Spatial Autoregressive model (SAR), Spatial Error Model (SEM), and Spatial Durbin Model (SDM). SAR incorporates the spatial lag term of the dependent variable, SEM includes that of the independent variables, while SDM encompasses all variables' spatial lag terms. Since EE is a restricted variable, the Tobit deformation of the above three models is used in this chapter. Tobit-SAR constructed in this chapter is as follows:

$$\mathrm{EE}_{it}^{*} = \alpha_0 + \rho W \times \mathrm{EE}_{it}^{*} + \beta \mathrm{GF}_{it} + \eta C_{it} + u_i + v_t + \varepsilon_{it} \tag{2.6}$$

where ρ captures the spatial spillover effects; the other symbols carry identical connotations as those in Eq. 2.5.

Tobit-SEM constructed in this chapter is as follows:

$$\begin{aligned} EE_{it} &= \alpha_0 + \beta GF_{it} + \eta C_{it} + u_i + v_t + \psi_{it} \\ \psi_{it} &= \lambda W \times \psi_{it} + \varepsilon_{it} \end{aligned} \tag{2.7}$$

where λ captures the spatial spillover effects; the other symbols carry identical connotations as those in Eq. 2.5.

Tobit-SDM constructed in this chapter is as follows:

$$\begin{aligned} EE_{it}^{*} = \alpha_0 + \rho W \times EE_{it}^{*} + \beta GF_{it} + \beta_1 W \times GF_{it} + \eta C_{it} \\ + \eta_1 W \times C_{it} + u_i + v_t + \varepsilon_{it} \end{aligned} \tag{2.8}$$

where ρ and β_1 captures the spatial spillover effects; the other symbols carry identical connotations as those in Eq. 2.5.

Analysis Results The results are in Table 2.7, where Column (1) is for the Tobit model, Columns (2)–(4) are for the Tobit-SAR, Columns (5)–(7) are for the Tobit-SEM, and Columns (8)–(10) are for the Tobit-SDM. The coefficient of GF based on the Tobit model is 0.033 at the 1% significant level. This finding means that GF contributes to enhancing EE, which confirms Hypothesis 2.2. The remaining columns are dedicated to examining the spatial spillover effects. These significant positive results of LR tests in all the columns evidence that these spatial models are superior to the Tobit model. Additionally, the ρ values or the λ values of these spatial models are all significantly positive. These findings collectively indicate the presence of spatial spillover effects, confirming Hypothesis 2.3.

As for the issue of model selection, the significant results of LM_Lag_Robust and LM_Error_Robust prove that SAR and SEM are superior to OLS model, respectively. What's more, the values of the Wald tests are significantly positive. It suggests that SDM is superior to SAR and SEM. All in all, Tobit-SDM is the most appropriate model for testing the impact of GF on EE. The results of Tobit-SDM in Columns (8)–(10) are consistent for all three weight matrices. And both GF and GF × **W** show significant positive coefficients, which proves the direct effects and the spatial spillover effects of GF on EE. Therefore, GF can contribute to the enhancement of local EE and, at the same time, promote the development of EE in neighboring cities. Specifically, the coefficient of GF × $\mathbf{W}_1$ is 0.007. It means that the scale of green bond issuance (in log, 100 million RMB) in neighboring cities, which are defined by their adjacent

Table 2.7 The results of effect of GF on EE

	Tobit	*Tobit-SAR*			*Tobit-SEM*			*Tobit-SDM*		
		$\boldsymbol{W}_1$	$\boldsymbol{W}_2$	$\boldsymbol{W}_3$	$\boldsymbol{W}_1$	$\boldsymbol{W}_2$	$\boldsymbol{W}_3$	$\boldsymbol{W}_1$	$\boldsymbol{W}_2$	$\boldsymbol{W}_3$
	(1)	*(2)*	*(3)*	*(4)*	*(5)*	*(6)*	*(7)*	*(8)*	*(9)*	*(10)*
GF	0.033*** (0.006)	0.042*** (0.011)	0.036*** (0.010)	0.034*** (0.010)	0.043*** (0.011)	0.034*** (0.011)	0.031*** (0.010)	0.039*** (0.010)	0.033*** (0.010)	0.030*** (0.009)
GF × $\mathbf{W}_1$								0.007** (0.003)		
GF × $\mathbf{W}_2$									0.046*** (0.017)	
GF × $\mathbf{W}_3$										0.039** (0.018)
Controls	√	√	√	√	√	√	√	√	√	√
ρ		0.039*** (0.009)	0.371*** (0.071)	0.408*** (0.067)				0.045*** (0.013)	0.328*** (0.083)	0.377*** (0.078)
λ					0.023*** (0.008)	0.379*** (0.083)	0.427*** (0.078)			
Sigma		0.210*** (0.014)	0.208*** (0.014)	0.205*** (0.014)	0.212*** (0.014)	0.209*** (0.014)	0.206*** (0.014)	0.208*** (0.015)	0.206*** (0.014)	0.204*** (0.014)
LM_Lag_Robust		10600***	593.534***	677.773***						
LM_Error_Robust					10400***	555.820***	632.484***			
Wald test								243.322***	308.127***	302.410***
LR test		19.581***	27.306***	37.085***	8.719***	20.984***	29.922***	12.058***	15.620***	23.373***
Observations	1692	1692	1692	1692	1692	1692	1692	1692	1692	1692
R^2		0.126	0.125	0.125	0.006	0.122	0.120	0.127	0.155	0.153

***$p < 0.01$. **$p < 0.05$. The values in parentheses are robust standard errors, clustered at the city level. The robust standard error of the Tobit model in Column (1) is obtained by bootstrap sampling 1000 times. $\mathbf{W}_1$, $\mathbf{W}_2$, and $\mathbf{W}_3$ represent the contiguity matrix, the economic matrix, and the nested matrix, respectively

boundaries with local cities, increases by one, local cities' energy efficiency can increase by 0.007. Other corresponding coefficients can be interpreted similarly.

This chapter, in conjunction with existing research, provides mutually reinforcing evidence. Utilizing provincial-level green credit data from China, Song et al. (2021) demonstrate that GF contributes to the enhancement of EE through positive spatial spillover effects. Similarly, Lee et al. (2023a) apply the entropy weighting method to analyze GF–EE nexus using Chinese provincial data. Their findings reveal that GF can significantly improve EE. Furthermore, Tan et al. (2025) find that green finance has improved corporate environmental performance both directly and indirectly by facilitating technological upgrades. In fact, GF has the potential to drive technological advancement, stimulate financial innovation, and strengthen environmental regulations, all of which contribute to increased urban EE. Furthermore, the economic activities among cities are often interdependent and interconnected. GF in one city can influence various economic factors in neighboring cities, such as production inputs, capital availability, and technological capabilities. These changes, in turn, affect EE in neighboring cities (Song et al., 2021).

2.5 Discussions

2.5.1 Contributions

The contributions are outlined here: First, it investigates the impact of GF at the city level. Cities represent the fundamental structure of modern society, and their sustainable development is crucial for the survival of humanity. Therefore, examining the role of GF in promoting urban sustainability is of paramount importance. Existing studies have primarily concentrated on national or provincial scales, posing challenges in isolating the precise impacts of GF at the urban level. Thus, this chapter enriches the body of research on GF. Second, it expands and advances the existing literature on the nexus between GF and AI. While there is substantial literature on the role of GF in driving technological innovation, there is a notable lack of studies exploring the correlation between GF and AI. Hence, we conduct a detailed examination of the relationship, thereby enhancing our understanding of the GF–AI nexus. Third, this chapter has evidenced the positive role of GF in enhancing EE and its spatial spillover effects. Limited research has been conducted on the potential of GF to improve EE, particularly regarding any spillover effects. This chapter provides a thorough analysis of the issue, filling a notable gap in the current literature.

2.5.2 *Policy Implications*

Firstly, it is crucial to actively accelerate the growth of GF. GF possesses the capability to allocate resources toward AI and energy sectors, thereby making a significant contribution to the advancement of both AI and EE. Consequently, it is imperative to expedite the enhancement of green financial system, foster green financial innovation, and strengthen oversight of GF to ensure its robust and sustainable development.

Secondly, it is crucial to develop GF-related policies tailored to local circumstances. It has been identified that how GF affects AI varies with various cities. In particular, GF is less effective in enhancing AI in western cities, resource-dependent cities, and cities with high levels of AI. For these cities, it is vital to reinforce supporting measures that effectively complement and enhance the functions of GF.

Thirdly, enhancing collaboration among cities is necessary to jointly promote improvements in EE and sustainable development. There exists a spatial spillover effect of GF on EE, which may lead cities to underestimate the significance of GF and become hesitant to pursue its development. To mitigate this issue, it is essential to strengthen regional cooperation, eliminate barriers to resource flow, and implement comprehensive stimulus policies for GF.

2.6 Conclusions

For an extended period, Chinese cities have pursued a model of urbanization characterized by high investment and low efficiency, which poses a significant threat to urban sustainable development. In recent years, GF has been anticipated to enhance urban environmental quality and promote sustainable development. However, there exists a significant gap in the research examining GF at the city level. To address this gap, this chapter utilizes 282 Chinese cities' datasets from 2014 to 2019 to investigate the effects of GF on AI and EE.

The results reveal that GF significantly contributes to the enhancement of AI levels in Chinese cities. An increase of 1 standard deviation in GF is associated with a rise of 0.043 standard deviations in the installation of industrial robots per thousand workers, as well as an increase of 0.062 standard deviations in the stock of industrial robots per thousand workers. These results remain robust when alternative measurements are employed, instrumental variable (IV) methods are applied, the influence of GF policies is controlled for, municipal samples are excluded, and DDL regression analyses are conducted. Furthermore, this chapter demonstrates that the beneficial influence of GF on AI is notably significant in eastern and central cities, non–resource-dependent cities, cities with a population under five million, and low-AI-level cities. Additionally, this research demonstrates that GF positively affects EE and exhibits a spatial spillover effect. That is, GF can contribute to the enhancement of local EE and, at the same time, promote the development of EE in neighboring cities.

Despite achieving significant research outcomes, the two inadequacies necessitate attention. Due to the restricted availability of city-level data, the analysis employs the issuance scale of green bonds as a proxy for GF. This approach may not allow for a comprehensive evaluation of GF levels within a city or an exploration of the heterogeneous impacts of various green financial instruments. Future research can build upon this topic once relevant city-level data becomes available. Furthermore, as this chapter does not focus on the nexus between AI and EE, empirical testing of their relationship is not conducted; instead, theoretical derivations are based on existing literature. Subsequent studies could further investigate the relationship between AI and EE.

References

Acemoglu, D., & Restrepo, P. (2020). Robots and jobs: Evidence from US labor markets. *Journal of Political Economy, 128*(6), 2188–2244. https://doi.org/10.1086/705716

Afzal, A., Rasoulinezhad, E., & Malik, Z. (2022). Green finance and sustainable development in Europe. *Economic Research-Ekonomska Istrazivanja, 35*(1), 5150–5163. https://doi.org/10.1080/1331677X.2021.2024081

Al-Mamun, M., Boubaker, S., & Nguyen, D. K. (2022). Green finance and decarbonization: Evidence from around the world. *Finance Research Letters, 46*, Article 102807. https://doi.org/10.1016/j.frl.2022.102807

Babina, T., Fedyk, A., He, A. L., & Hodson, J. (2024). Artificial intelligence, firm growth, and product innovation. *Journal of Financial Economics, 151*, Article 103745. https://doi.org/10.1016/j.jfineco.2023.103745

Bai, R., Wu, H. X., Tan, Z. Z., & Hong, T. (2025). Green finance and green innovation: The moderating role of ESG and synergies with inclusive finance. *Research in International Business and Finance, 79*, Article 103056. https://doi.org/10.1016/j.ribaf.2025.103056

Bakry, W., Mallik, G., Nghiem, X. H., Sinha, A., & Vo, X. V. (2023). Is green finance really green? Examining the long-run relationship between green finance, renewable energy and environmental performance in developing countries. *Renewable Energy, 208*, 341–355. https://doi.org/10.1016/j.renene.2023.03.020

Chen, L. F., & Wang, K. F. (2022). The spatial spillover effect of low-carbon city pilot scheme on green efficiency in Chinese cities: Evidence from a quasi-natural experiment. *Energy Economics, 110*, Article 106018. https://doi.org/10.1016/j.eneco.2022.106018

Chen, J., Meng, W. F., Chen, Y. L., & Zhou, W. (2024). To be an eco- and tech-friendly society: Impact research of green finance on AI innovation. *Journal of Cleaner Production, 466*, Article 142900. https://doi.org/10.1016/j.jclepro.2024.142900

Cui, Y. N., Zhong, C., Cao, J. H., & Guo, M. Y. (2023). Can green finance effectively mitigate $PM_{2.5}$ pollution? What role will green technological innovation play? *Energy & Environment*, Article 0958305X231204030. https://doi.org/10.1177/0958305X231204030

Debrah, C., Chan, A. P. C., & Darko, A. (2022). Green finance gap in green buildings: A scoping review and future research needs. *Building and Environment, 207*, Article 108443. https://doi.org/10.1016/j.buildenv.2021.108443

Fingleton, B., & Gallo, J. L. (2008). Estimating spatial models with endogenous variables, a spatial lag and spatially dependent disturbances: Finite sample properties. *Papers in Regional Science, 87*(3), 319–340. https://doi.org/10.1111/j.1435-5957.2008.00187.x

Guo, Z., Ćevid, D., & Buhlmann, P. (2022). Doubly Debiased LASSO: High-dimensional inference under hidden confounding. *The Annals of Statistics, 50*(3), Article 1320. https://doi.org/10.1214/21-AOS2152

Guo, Q. T., Dong, Y., Feng, B., & Zhang, H. (2023). Can green finance development promote total-factor energy efficiency? Empirical evidence from China based on a spatial Durbin model. *Energy Policy, 177*, Article 113523. https://doi.org/10.1016/j.enpol.2023.113523

Hong, Q. Q., Cui, L. H., & Hong, P. H. (2022). The impact of carbon emissions trading on energy efficiency: Evidence from quasi-experiment in China's carbon emissions trading pilot. *Energy Economics, 110*, Article 106025. https://doi.org/10.1016/j.eneco.2022.106025

Hu, J. L., & Wang, S. C. (2006). Total-factor energy efficiency of regions in China. *Energy Policy, 34*(17), 3206–3217. https://doi.org/10.1016/j.enpol.2005.06.015

Huang, H. Y., Mbanyele, W., & Wang, Y. Z. (2022a). Climbing the quality ladder of green innovation: Does green finance matter? *Technological Forecasting and Social Change, 184*, Article 122007. https://doi.org/10.1016/j.techfore.2022.122007

Huang, Y. M., Chen, C., Lei, L. J., & Zhang, Y. P. (2022b). Impacts of green finance on green innovation: A spatial and nonlinear perspective. *Journal of Cleaner Production, 365*, Article 132548. https://doi.org/10.1016/j.jclepro.2022.132548

Huo, D., Zhang, X. T., Meng, S., Wu, G., Li, J. H., & Di, R. Q. (2022). Green finance and energy efficiency: Dynamic study of the spatial externality of institutional support in a digital economy by using hidden Markov chain. *Energy Economics, 116*, Article 106431. https://doi.org/10.1016/j.eneco.2022.106431

Lee, C. C., & Hussain, J. (2023). Energy sustainability under the COVID-19 outbreak: Electricity break-off policy to minimize electricity market crises. *Energy Economics, 125*, Article 106870. https://doi.org/10.1016/j.eneco.2023.106870

Lee, C. C., & Lee, C. C. (2022). How does green finance affect green total factor productivity? Evidence from China. *Energy Economics, 107*, Article 105863. https://doi.org/10.1016/j.eneco.2022.105863

Lee, C. C., Wang, C. S., He, Z. W., Xing, W. W., & Wang, K. Y. (2023a). How does green finance affect energy efficiency? The role of green technology innovation and energy structure. *Renewable Energy, 219*(1), Article 119417. https://doi.org/10.1016/j.renene.2023.119417

Lee, C. C., Wang, F. H., & Chang, Y. F. (2023b). Does green finance promote renewable energy? Evidence from China. *Resources Policy, 82*, Article 103439. https://doi.org/10.1016/j.resourpol.2023.103439

Lee, C. C., Wang, F. H., & Chang, Y. F. (2023c). Towards net-zero emissions: Can green bond policy promote green innovation and green space? *Energy Economics, 121*, Article 106675. https://doi.org/10.1016/j.eneco.2023.106675

Lei, H., Gao, R. Y., Ning, C. T., & Sun, G. L. (2025). Green finance and corporate green innovation. *Finance Research Letters, 72*, Article 106577. https://doi.org/10.1016/j.frl.2024.106577

Li, J., Ma, S. Q., Qu, Y., & Wang, J. M. (2023). The impact of artificial intelligence on firms' energy and resource efficiency: Empirical evidence from China. *Resources Policy, 82*, Article 103507. https://doi.org/10.1016/j.resourpol.2023.103507

Li, X. Y., Wang, Q., & Tang, Y. (2024). The impact of artificial intelligence development on urban energy efficiency—Based on the perspective of smart city policy. *Sustainability, 16*(8), Article 3200. https://doi.org/10.3390/su16083200

Li, Q., Lei, H. Y., & Yu, M. G. (2025). Green finance, green innovation and carbon intensity. *International Review of Financial Analysis, 106*, Article 104550. https://doi.org/10.1016/j.irfa.2025.104550

Liao, N., & He, Y. (2018). Exploring the effects of influencing factors on energy efficiency in industrial sector using cluster analysis and panel regression model. *Energy, 158*, 782–795. https://doi.org/10.1016/j.energy.2018.06.049

Liu, L., Yang, K., Fujii, H., & Liu, J. (2021). Artificial intelligence and energy intensity in China's industrial sector: Effect and transmission channel. *Economic Analysis and Policy, 70*, 276–293. https://doi.org/10.1016/j.eap.2021.03.002

Liu, H. D., Yao, P. B., Latif, S., Aslam, S., & Iqbal, N. (2022). Impact of green financing, FinTech, and financial inclusion on energy efficiency. *Environmental Science and Pollution Research, 29*(13), 18955–18966. https://doi.org/10.1007/s11356-021-16949-x

Liu, Y., Xie, Y. X., & Zhong, K. Y. (2024). Impact of digital economy on urban sustainable development: Evidence from Chinese cities. *Sustainable Development, 32*(1), 307–324. https://doi.org/10.1002/sd.2656

Luo, Q. F., & Feng, P. C. (2024). Exploring artificial intelligence and urban pollution emissions: "Speed bump" or "accelerator" for sustainable development? *Journal of Cleaner Production, 463*, Article 142739. https://doi.org/10.1016/j.jclepro.2024.142739

Lv, C. C., Bian, B. C., Lee, C. C., & He, Z. W. (2021). Regional gap and the trend of green finance development in China. *Energy Economics, 102*, Article 105476. https://doi.org/10.1016/j.eneco.2021.105476

Ma, L. D., Hong, Y. X., He, S. C., Luo, H. B., Liu, G. M., Zheng, J. H., Xia, Y. T., & Xiao, D. H. (2023). The influence of digital economy development on urban carbon emission intensity in the Yangtze River Economic Belt: Mediating mechanism and spatial effect. *Frontiers in Ecology and Evolution, 11*, Article 1148505. https://doi.org/10.3389/fevo.2023.1148505

Ma, Z. X., Pu, D. P., & Liang, H. J. (2024). Financing net-zero energy integration in smart cities with green bonds and public-private partnerships. *Sustainable Energy Technologies and Assessments, 64*, Article 103708. https://doi.org/10.1016/j.seta.2024.103708

Madaleno, M., Dogan, E., & Taskin, D. (2022). A step forward on sustainability: The nexus of environmental responsibility, green technology, clean energy and green finance. *Energy Economics, 109*, Article 105945. https://doi.org/10.1016/j.eneco.2022.105945

Naeem, M. A., Conlon, T., & Cotter, J. (2021). Green bonds and other assets: Evidence from extreme risk transmission. *Journal of Environmental Management, 305*, Article 114358. https://doi.org/10.1016/j.jenvman.2021.114358

Qu, C. Y., Shao, J., & Shi, Z. K. (2020). Does financial agglomeration promote the increase of energy efficiency in China? *Energy Policy, 146*, Article 111810. https://doi.org/10.1016/j.enpol.2020.111810

Shu, Y. X., Deng, N. X., Wu, Y. M., Bao, S. M., & Bie, A. (2023). Urban governance and sustainable development: The effect of smart city on carbon emission in China. *Technological Forecasting and Social Change, 193*, Article 122643. https://doi.org/10.1016/j.techfore.2023.122643

Song, M. L., Xie, Q. J., & Shen, Z. Y. (2021). Impact of green credit on high-efficiency utilization of energy in China considering environmental constraints. *Energy Policy, 153*, Article 112267. https://doi.org/10.1016/j.enpol.2021.112267

Song, A. F., Rasool, Z., Nazar, R., & Anser, M. K. (2024). Towards a greener future: How green technology innovation and energy efficiency are transforming sustainability. *Energy, 290*, Article 129891. https://doi.org/10.1016/j.energy.2023.129891

Su, C. W., Umar, M., & Gao, R. S. (2022). Save the environment, get financing! How China is protecting the environment with green credit policies? *Journal of Environmental Management, 323*, Article 116178. https://doi.org/10.1016/j.jenvman.2022.116178

Sun, H. P., Edziah, B. K., Sun, C. W., & Kporsu, A. K. (2019). Institutional quality, green innovation and energy efficiency. *Energy Policy, 135*, Article 111002. https://doi.org/10.1016/j.enpol.2019.111002

Tan, Y. X., Lin, B., & Wang, L. Y. (2025). Green finance and corporate environmental performance. *International Review of Economics & Finance, 98*, Article 103929. https://doi.org/10.1016/j.iref.2025.103929

Wan, G. C., Zhang, W. K., & Li, C. (2024). How does low-carbon city pilot policy catalyze companies toward ESG practices? Evidence from China. *Economic Analysis and Policy, 81*, 1593–1607. https://doi.org/10.1016/j.eap.2024.02.036

Wang, L. H., & Shao, J. (2023). Digital economy, entrepreneurship and energy efficiency. *Energy Economics, 269*, Article 126801. https://doi.org/10.1016/j.energy.2023.126801

Wang, X. Y., & Wang, Q. (2021). Research on the impact of green finance on the upgrading of China's regional industrial structure from the perspective of sustainable development. *Resources Policy, 74*, Article 102436. https://doi.org/10.1016/j.resourpol.2021.102436

Wang, Z. L., & Wang, X. B. (2022). Research on the impact of green finance on energy efficiency in different regions of China based on the DEA-Tobit model.

Resources Policy, 77, Article 102695. https://doi.org/10.1016/j.resourpol.2022.102695

Wang, E. Z., Lee, C. C., & Li, Y. Y. (2022). Assessing the impact of industrial robots on manufacturing energy intensity in 38 countries. *Energy Economics, 105*, Article 105748. https://doi.org/10.1016/j.eneco.2021.105748

Wang, P. H., Song, W., Zhou, J. L., Tan, Y. S., & Wang, H. K. (2023). AI-based environmental color system in achieving sustainable urban development. *Systems, 11*(3), Article 135. https://doi.org/10.3390/systems11030135

Wei, X. H., Zhang, P. K., & Du, Y. H. (2020). How robots reshape the urban labor market: From a perspective of migrants' job tasks. *Economic Perspective*, 92–109. In Chinese.

Wei, G. E., He, B. J., Sun, P. J., Liu, Y. B., Li, R. Z., Ouyang, X., Luo, K., & Li, S. S. (2023). Evolutionary trends of urban expansion and its sustainable development: Evidence from 80 representative cities in the belt and road initiative region. *Cities, 138*, 104353. https://doi.org/10.1016/j.cities.2023.104353

Wei, Z., Iqbal, A., Jahangir, S., Ali, M. S. E., & Hussain, M. (2025). Financing the green transition: How green finance, green innovation, green growth, and environmental taxes can drive carbon neutrality. *Environment, Development and Sustainability*. https://doi.org/10.1007/s10668-024-05933-3.

Wu, J., Lin, K. X., & Sun, J. S. (2023). Improving urban energy efficiency: What role does the digital economy play? *Journal of Cleaner Production, 418*, Article 138104. https://doi.org/10.1016/j.jclepro.2023.138104

Xu, A. T., Zhu, Y. H., & Wang, W. P. (2023). Micro green technology innovation effects of green finance pilot policy—From the perspectives of action points and green value. *Journal of Business Research, 159*, 113724. https://doi.org/10.1016/j.jbusres.2023.113724

Yang, L. S., & Ni, M. Y. (2022). Is financial development beneficial to improve the efficiency of green development? Evidence from the "belt and road" countries. *Energy Economics, 105*, Article 105734. https://doi.org/10.1016/j.eneco.2021.105734

Yin, K. D., Cai, F. F., & Huang, C. (2023). How does artificial intelligence development affect green technology. *Environmental Science and Pollution Research, 30*, 28066–28090. https://doi.org/10.1007/s11356-022-24088-0

You, J. S., Hu, J., & Jiang, B. (2025). Non-stationarity and spatial spillover effects in artificial intelligence development: Implications for sustainable urban transformation. *Sustainable Cities and Society, 131*, Article 106746. https://doi.org/10.1016/j.scs.2025.106746

Yu, C. H., Wu, X. Q., Zhang, D. Y., Chen, S., & Zhao, J. S. (2021). Demand for green finance: Resolving financing constraints on green innovation in China. *Energy Policy, 153*, Article 112255. https://doi.org/10.1016/j.enpol.2021.112255

Yue, H., Zhou, Z. Z., & Liu, H. W. (2024). How does green finance influence industrial green total factor productivity? Empirical research from China. *Energy Reports, 11*, 914–924. https://doi.org/10.1016/j.egyr.2023.12.056

Zeba, G., Dabic, M., Cicak, M., Daim, T., & Yalcin, H. (2021). Technology mining: Artificial intelligence in manufacturing. *Technological Forecasting and Social Change, 171*, Article 120971. https://doi.org/10.1016/j.techfore.2021.120971

Zeng, M., & Zhang, W. K. (2024). Green finance: The catalyst for artificial intelligence and energy efficiency in Chinese urban sustainable development. *Energy Economics, 139*, Article 107883. https://doi.org/10.1016/j.eneco.2024.107883

Zhang, G. W. (2023). The heterogeneous role of green finance on industrial structure upgrading – Based on spatial spillover perspective. *Finance Research Letters, 58*, Article 104596. https://doi.org/10.1016/j.frl.2023.104596

Zhang, W. K., & Zeng, M. (2024). Is artificial intelligence a curse or a blessing for enterprise energy intensity? Evidence from China. *Energy Economics, 134*, Article 107561. https://doi.org/10.1016/j.eneco.2024.107561

Zhang, W. K., Fan, H. X., & Zhao, Q. W. (2023a). Seeing green: How does digital infrastructure affect carbon emission intensity? *Energy Economics, 127*, Article 107085. https://doi.org/10.1016/j.eneco.2023.107085

Zhang, W. K., Luo, Q., Zhang, Y. F., & Yu, A. (2023b). Does green credit policy matter for corporate exploratory innovation? Evidence from Chinese enterprises. *Economic Analysis and Policy, 80*, 820–834. https://doi.org/10.1016/j.eap.2023.09.024

Zhang, W. K., Zeng, M., Zhang, Y. F., & Su, C. W. (2023c). Reducing carbon emissions: Can high-speed railway contribute? *Journal of Cleaner Production, 413*, Article 137524. https://doi.org/10.1016/j.jclepro.2023.137524

Zhang, M. L., Zheng, L., & Xie, Y. (2023d). Spatial spillover effect of green finance and clean energy on China's provincial green economic development. *Environmental Science and Pollution Research, 30*(29), 74070–74092. https://doi.org/10.1007/s11356-023-27450-y

Zhang, W. K., Fan, H. X., & Zeng, M. (2025). Unleashing the power of artificial intelligence: A game changer for urban energy efficiency in China. *Sustainable Cities and Society, 126*, Article 106372. https://doi.org/10.1016/j.scs.2025.106372

Zhao, H., Chen, S. W., & Zhang, W. K. (2023). Does digital inclusive finance affect urban carbon emission intensity: Evidence from 285 cities in China. *Cities, 142*(11), Article 104552. https://doi.org/10.1016/j.cities.2023.104552

Zhou, P., Ang, B. W., & Wang, H. (2012). Energy and CO_2 emission performance in electricity generation: A non-radial directional distance function approach. *European Journal of Operational Research, 221*, 625–635. https://doi.org/10.1016/j.ejor.2012.04.022

Zhu, X. D., Zhou, X. Y., Zhu, S. J., & Huang, H. F. (2021). Urban green finance in China and its influencing factors: A case study of green bonds. *Journal of Natural Resources., 36*(12), 3247–3260. In Chinese.

CHAPTER 3

AI-Driven Urban Energy Efficiency Revolution

3.1 Introduction

The enhancement of energy security and the promotion of energy sustainability have become significant topics of inquiry within both academic and governmental domains (Lee & Hussain, 2023; Zhang et al., 2025). As global energy consumption continues to rise, the challenge of energy scarcity has increasingly come to the forefront (Su et al., 2022; Zhang et al., 2025). China continues to exhibit high levels of energy consumption (Chen et al., 2024; Zhang et al., 2025). For instance, in 2022, its primary energy consumption reached 159.39 exajoules, representing 26.4% of global consumption. This figure reflects an average growth rate of 3.1% from 2012 to 2022, as demonstrated in Fig. 3.1. This growth rate significantly surpasses the global average of 1.4%.[1] Urban areas in China are particularly significant in terms of energy consumption, accounting for approximately 85% of the nation's total, in contrast to a global average of 67%. Nevertheless, the energy efficiency of Chinese cities is suboptimal, as

[1] 1 exajoule ≈ 23.88 million tons of oil equivalent.

The core content of this chapter is derived from this article: Zhang, W. K., Fan, H. X., & Zeng, M. (2025). Unleashing the Power of Artificial Intelligence: A Game Changer for Urban Energy Efficiency in China. *Sustainable Cities and Society*, *126*, Article 106372. https://doi.org/10.1016/j.scs.2025.106372

W. Zhang et al., *Green Transformation in the Digital Age*,
https://doi.org/10.1007/978-981-95-9121-3_3

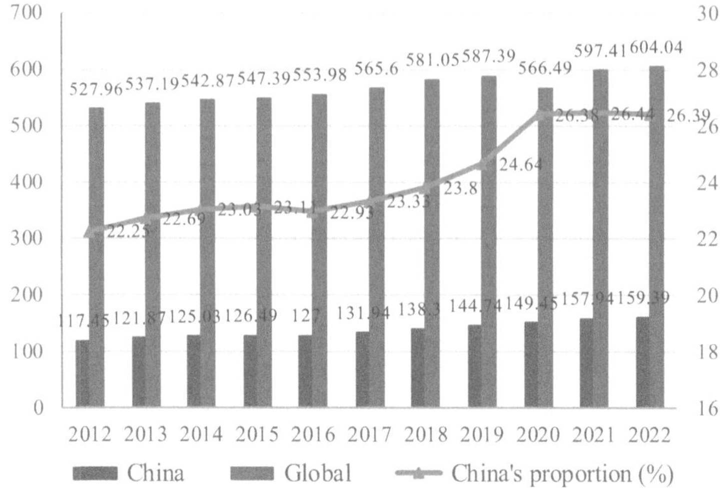

Fig. 3.1 Annual primary energy consumption (exajoule) during 2012–2022

the energy intensity per unit of GDP in these urban centers is approximately 1.4 times higher than the global average.[2] Therefore, improving urban energy efficiency (EE) in China is crucial for alleviating energy pressures and fostering sustainable energy practices (Wei et al., 2023; Zhang et al., 2025).

In recent years, artificial intelligence (AI) has entered a significant phase of development, driven by advancements in technologies such as big data analytics, sensor networks, and neuroscience (Acemoglu & Restrepo, 2020). This evolution has led to the deep integration of AI across various sectors of the economy and society. Nations around the globe are engaged in intense competition to enhance their AI capabilities and secure a competitive advantage in this critical area. In 2017, China introduced the New Generation of Artificial Intelligence Development Plan, which aims to leverage substantial strategic opportunities in AI and accelerate the country's transformation into an innovative nation and a global leader in science and technology. Following years of progress, China has made notable strides in AI development. For instance, the installation of industrial robots in China exceeded 290,000 units in 2022, with the total number

[2] People's Daily: http://paper.people.com.cn

of robots in the country surpassing 1.5 million units, representing approximately 52% and 38% of the global totals, respectively.[3]

The utilization of AI has yielded significant accomplishments in various domains, including technological innovation (Jung & Lim, 2020), the stimulation of economic growth (Berg et al., 2018), and the improvement of individual well-being (Makridis & Mishra, 2022). Furthermore, AI exhibits considerable promise in fostering sustainable urban development (Allam & Dhunny, 2019; Leal et al., 2024; You et al., 2025). Nevertheless, the implications of AI for urban EE remain ambiguous due to its dual impact on energy consumption. On the one hand, AI promotes the development of environmentally friendly technologies and optimizes industrial frameworks, which effectively reduces urban energy consumption and lessens pollution emissions in urban areas (Benzidia et al., 2021; Mehmood et al., 2019; Luo et al., 2025). Conversely, the deployment of AI requires substantial data processing and training, resulting in significant energy consumption (Dauvergne, 2022; Luo et al., 2023). Furthermore, the existing literature on urban EE predominantly emphasizes aspects of the digital economy (e.g., Zhao et al., 2023; Wu et al., 2023), public energy policy (e.g., Gao et al., 2022a), and urban governance (e.g., Khavarian-Garmsir et al., 2023; Wang et al., 2019), with scant attention given to the influence of AI (Li et al., 2024). Therefore, while these studies enhance the understanding of the relationship between AI and urban EE, there remains a pressing need for direct investigation into the effects of AI on urban EE.

This chapter investigates the effect of AI on urban EE employing data from 282 Chinese cities spanning the years 2006–2019. Initially, we examine how AI affects urban EE, and the findings evidence that AI benefits the enhancement of urban EE. The validity of these conclusions remains unchanged when adopting alternative measurements, removing the municipality samples, and using other methods including the panel Tobit model, DDL regression, and instrumental variable (IV) method. Furthermore, we analyze the heterogeneity of AI's effects on urban EE across different city typologies. The results demonstrate that the enhancement effect of urban EE brought by AI is further accentuated in resource-dependent cities, eastern and central cities, northern cities of the Qinling Mountains–Huaihe River Line, and mega-sized and super-sized cities. We finally analyze the AI's spatial spillover effect on urban EE leveraging the

[3] IFR: https://ifr.org/

Spatial Durbin Model (SDM). The results indicate that AI exerts positive spatial spillover effects on urban EE.

3.2 Theoretical Analysis

3.2.1 Literature Review

The swift advancement of AI has garnered significant interest within the academic community regarding its implications across various domains, including economic, social, and cultural spheres. In light of the focus of this chapter, we aim to conduct a comprehensive review of the existing literature concerning the influence of AI on sustainable development, with particular emphasis on urban sustainable development. AI has exhibited remarkable capabilities in advancing sustainable development and possesses significant potential to enhance the sustainability of urban environments and communities (Anthopoulos & Kazantzi, 2022).

Certain scholarly works examine the influence of AI on urban carbon emissions (Ding et al., 2023; Liu et al., 2022; Wang et al., 2024). Zhang et al. (2024) employed the Attention Deep and Cross Network model to assess the effects of AI technology on energy consumption and carbon emissions within Chinese cities. Their findings suggest that the implementation of AI could enhance the reduction of carbon emissions in these urban areas by an estimated 1 quintillion tons. In a similar vein, Dong et al. (2024) examined the correlation between AI and urban carbon emissions, revealing that the influence of various AI types is not uniform. Their findings indicate that applied AI has the potential to decrease urban carbon emissions; however, the effects of innovative AI on these emissions are multifaceted. Specifically, innovative AI initially contributes to an increase in urban carbon emissions, which is subsequently followed by a gradual reduction. Certain scholarly works examine the influence of AI on greening of urban infrastructure (Yigitcanlar et al., 2020). Son et al. (2023) examined the contribution of AI to urban planning, emphasizing its function in enhancing the technological capabilities of local governments and in fostering collaboration and communication among various stakeholders. This analysis delineates the significance of AI in advancing smart and sustainable urban development. Regona et al. (2024) posit that AI has the potential to enhance various phases of urban building processes, including planning, construction, operation, and maintenance. Furthermore, they contend that AI could fundamentally transform the urban construction sector, steering it toward greater efficiency, sustainability, and environmental responsibility. Certain scholarly works examine the influence of AI on urban governance (Sepehr, 2024). Utilizing advancements in AI, Chang et al. (2024) employed a dynamic Peephole LSTM model to forecast urban power consumption and demand, which aims to enhance urban power management, facilitate the modernization of urban energy systems, and bolster the capabilities of urban governance. Bibri et al. (2024) investigated the underlying principles and operational frameworks of Artificial Intelligence of Things (AIoT) within the context of urban governance, with a particular focus on environmental governance. Their findings suggest that AIoT serves as a foundational element for a novel model of data-driven urban governance.

Current research has also examined the influencing factors of urban EE from multiple perspectives, encompassing, but not restricted to, technological advancements, environmental regulations, resource distribution, and industrial organization (Cui & Cao, 2024; Song et al., 2023). This chapter will exclusively focus on technological advancements, in accordance with its subject matter. A substantial body of research has established the significance of technological advancements in enhancing energy efficiency (Tian et al., 2024). For example, utilizing the strategic initiative known as "Broadband China," Ma and Lin (2023) conducted an empirical analysis to assess the influence of digital infrastructure development on urban EE. Their findings indicate that the digitization of infrastructure significantly enhances the efficiency of energy utilization in urban settings. Similarly, Gao et al. (2022b) evidence that urban EE has experienced substantial enhancement as a result of advancements in information and communication technology (ICT) within Chinese urban areas. Furthermore, Zhang and Fu (2022) found that advancements in technology, whether resulting from independent innovation or external sources, can significantly enhance urban EE.

AI, as a state-of-the-art digital technology, serves as a significant indicator of technological advancement in the modern age. Consequently, the impact of AI on energy efficiency warrants careful consideration, a topic that has garnered interest in various scholarly works. Zhou et al. (2024) conducted an empirical analysis utilizing provincial-level data from China, revealing that advancements in AI have facilitated energy conservation and emission reduction within the country. Similarly, utilizing provincial, industry, and enterprise-specific data, respectively, Li et al. (2025), Liu et al. (2021), and Song et al. (2024) have illustrated that AI contributes to enhancements in energy efficiency.

The existing literature contains a substantial body of research on AI and urban EE, yielding significant findings. However, there is a paucity of studies that have directly investigated the influence of AI on urban EE. Theoretically, AI has the potential to enhance energy conservation through advancements in technology; conversely, the operation of AI systems is associated with considerable energy consumption. This duality complicates the assessment of AI's overall impact on urban EE. Our review identified only one empirical study that specifically analyzed the relationship between AI and urban EE. Li et al. (2024) find that AI helps improve urban EE using data of Chinese cities; however, they fail to investigate the potential variability in the effects of AI on urban EE across diverse urban settings. Therefore, it is evident that further investigation into the AI-urban EE relationship is warranted.

3.2.2 *The Effect of AI on Urban EE*

Energy efficiency serves as a critical metric for assessing the advantages associated with unit energy consumption, while urban EE examines the overall energy efficiency performance at the city level. Numerous factors influence urban EE, which include not only various elements within the urban environment but also the structural interconnections and combinations among these elements. Consequently, the influence of AI on urban EE is complex and multifaceted (Sharifi et al., 2024; Zeng & Zhang, 2024).

At the micro-level, AI has the potential to enhance urban EE by improving the energy efficiency of enterprises, which are the primary agents of urban production activities. AI facilitates technological advancements within these enterprises, encourages the adoption of environmentally sustainable technologies, and, consequently, increases their green total factor productivity and energy efficiency (Liu et al., 2021; Yin et al., 2023). Furthermore, AI can effectively address the information asymmetry between enterprises and financial institutions, thereby aiding enterprises in securing green credit and improving their energy utilization efficiency (Lee & Lee, 2022). AI also supports energy conservation efforts by continuously monitoring real-time energy consumption and providing timely optimization recommendations (Ahmad et al., 2021). In addition, AI can play a significant role in enhancing urban EE by encouraging urban residents to adopt environmentally friendly lifestyles (Zhang & He, 2020).

For instance, AI-driven smart travel route planning can alleviate traffic congestion by directing residents away from congested routes, resulting in decreased energy consumption and lower pollution emissions (Wong et al., 2023). Similarly, the rise of e-commerce among residents has been significantly influenced by AI's effectiveness in optimizing logistics routes, thereby reducing the necessity for physical travel and, in turn, decreasing associated energy consumption (Zhang et al., 2020).

At the meso-level, AI has the potential to enhance urban EE by facilitating the green transformation of various urban industries. A multitude of studies has highlighted the promising role of AI in advancing environmentally sustainable, low-carbon, and highly efficient energy sectors (Maleki et al., 2022). In particular, AI-driven smart grids exhibit significant potential for improving electricity utilization efficiency. Advanced AI methodologies have demonstrated effectiveness in optimizing electric power consumption, which is essential for achieving urban energy conservation and emission reduction objectives (Boza & Evgeniou, 2021; Ounifi et al., 2022). Moreover, AI has made considerable contributions to the development and implementation of diverse renewable energy sources (Urom et al., 2022). AI also plays a crucial role in promoting the proliferation of green transportation alternatives, such as public transit systems, bike-sharing initiatives, and electric vehicles. Specifically, AI can intelligently manage traffic signals, optimize public transportation timetables, streamline bike borrowing and returning processes via mobile applications, and provide technical support for electric vehicle charging (Zulfiqar et al., 2023). Furthermore, AI significantly contributes to the advancement of green building practices. It can enhance building design through sophisticated algorithms, offer substantial support for building automation systems, and accurately predict and monitor energy consumption within buildings (Chen, 2015). The integration of AI in the construction, operation, and maintenance of buildings has been widely adopted to effectively reduce energy consumption associated with these activities.

At the macro-level, AI has the potential to enhance urban EE by optimizing the urban economic structure, transforming economic models, and facilitating the development of a green urban economy (Chang et al., 2023). AI plays a crucial role in the emergence and evolution of the digital economy, which is characterized by a data-driven approach that minimizes energy consumption and emissions (Pan & Nishant, 2023). Indeed, AI serves as the cornerstone of the digital economy, enabling the efficient collection, filtering, storage, and utilization of large datasets, thereby facilitating optimal resource allocation and productivity enhancement. Furthermore, AI contributes to the advancement of the urban economy by facilitating the reallocation of resources from low-efficiency sectors to those with higher efficiency. For instance, research by Zou and Xiong (2023) demonstrates that AI has enabled the systematic phasing out of energy-intensive industries, redirecting resources toward cleaner sectors, such as those that are technology- and knowledge-intensive, thus achieving industrial upgrading and rationalization. Additionally, AI supports efficient urban energy planning and resource optimization by fostering industrial integration. By leveraging advanced digital technologies, AI compresses time and space, thereby enhancing inter-sectoral collaboration and reinforcing the momentum of high-efficiency sectors (Chen & Liu, 2021).

Therefore, we develop the following hypothesis:

Hypothesis 3.1: AI can significantly enhance urban EE.

3.2.3 The Spatial Spillover Effect of AI on Urban EE

Local urban EE is shaped not only by AI resources within a specific city but also by those in surrounding regions (Ma et al., 2023; Zhang et al., 2023a). Consequently, improvements in local urban EE resulting from advancements in AI can extend benefits to other cities within the larger industrial network. The interconnectedness of these urban areas enables them to leverage the AI applications of their neighbors, thereby promoting a collective enhancement of urban EE. Moreover, cities that are leaders in AI technology are likely to exert a spillover effect on adjacent cities, driving technological progress in those areas, a phenomenon referred to as spatial diffusion (Tao et al., 2023). Additionally, competition among cities may incentivize those that are lagging to acquire knowledge from their more advanced counterparts in the AI sector, thus facilitating a continuous improvement in their technological capabilities (Yu et al., 2016). The substantial advantages that local AI provides to neighboring cities play a crucial role in enhancing their urban EE. Furthermore, the emergence of AI transcends geographical limitations, enabling the compression of time and space, which in turn promotes improved knowledge exchange, technological innovation, and collaboration among talent (Du et al., 2023). Therefore, the deployment of AI serves to optimize and integrate resources across various cities, fostering economic synergies and enhancing their urban EE. In summary, the impact of AI on urban EE demonstrates a spatial spillover effect, indicating that AI can improve urban EE not only within local cities but also in their neighboring cities.

Therefore, we develop the following hypothesis:

Hypothesis 3.2: AI can exert a significant spatial spillover effect on urban EE enhancement.

3.3 Research Design

3.3.1 Data

By leveraging the data of Chinese cities spanning the years 2006–2019, we investigate the influence of AI on EE. The data can be categorized into two main types. The first category comprises AI-related data, specifically the prevalence of industrial robots, sourced from the International Federation of Robotics (IFR). The IFR offers information on industrial robots in China; however, this data is reported at the industry level rather than at city level, necessitating a decomposition of the data for our analysis. The second category consists of city-level data, which is obtained from the CNRDS.[4]

3.3.2 Variable

The Explained Variable

Currently, two methodologies are employed to measure urban EE (Hong et al., 2022). The first method, known as the single-factor approach, evaluates urban EE by analyzing urban economic output per unit of energy input. However, this approach has limitations, as it focuses exclusively on energy consumption and ignores the contribution of various other inputs to production activities. Additionally, it considers economic output as the sole outcome, overlooking the emissions of pollutants associated with production. The total-factor approach offers a more holistic analysis, making it a scientifically superior method (Sun et al., 2019). Therefore, we adopt the total-factor approach to measure urban EE.

[4] Chinese Research Data Services Platform (CNRDS): https://www.cnrds.com/

This chapter incorporates three input factors: aggregate energy consumption, fixed capital stock, and labor force (Chen & Wang, 2022; Qu et al., 2020). On the output side, we analyze both desired and undesired outputs. The desired output is represented by urban real GDP, while the undesired outputs include urban pollutant emissions such as industrial soot, sulfur dioxide, and effluent.

Based on the above multiple inputs and outputs, this chapter leverages the super-efficiency Data Envelopment Analysis (DEA) method to calculate Chinese urban EE. This method possesses many advantages, including the absence of uniform variable dimensions and the avoidance of weight assumptions. Meanwhile, by introducing slack variables, the super-efficiency DEA model addresses the limitation of traditional DEA by enabling further sorting of effective units, thereby avoiding the problem of slack bias (Zhou et al., 2012). This chapter adopts an input-oriented radial distance function and assumes constant returns to scale, following Zhang and Zeng, (2024). The detailed calculation is set as follows:

$$\min\left[\theta-\varepsilon\left(\sum_{i=1}^{m}s_i^-+\sum_{r=1}^{s}s_r^+\right)\right]$$

$$s.t.\begin{cases}\sum\limits_{j=1,j\neq j_0}^{n}X_{ij}\lambda_j+s_i^-=\theta X_{ij_0},i=1,2,\ldots,m\\ \sum\limits_{j=1,j\neq j_0}^{n}Y_{rj}\lambda_j-s_r^+=Y_{rj_0},r=1,2,\ldots,s\\ \lambda_j,s_i^-,s_r^+\geq 0,j=1,2,\ldots,n\end{cases}\tag{3.1}$$

where X indicates the input vector, and Y indicates the output vector; θ captures urban EE of city j_0; λ_j defines the weight coefficients of the j th city; and s_i^- and s_r^+ are slack variables measuring the differences between the actual values and the optimal values of inputs and outputs.

The Explanatory Variable

We utilize exposure to industrial robots to assess AI. An industrial robot is defined as a device that can be automatically controlled and programmed to replicate human intelligence, representing a key form of AI (Acemoglu & Restrepo, 2020). The exposure to

(continued)

(continued)

industrial robots quantifies the extent of industrial robot implementation, which is calculated by the number of industrial robots per unit of employment. Referring to Wei et al. (2020) and Zhang & Zeng, (2024), we derive urban exposure to industrial robots by breaking down the exposure at industry level, as illustrated in Eq. 3.2:

$$AI_{it} = \sum_{j=1}^{J} \frac{empl_{i,j,t=2005}}{empl_{i,t=2005}} \times \frac{robot_{j,t}}{empl_{j,t=2005}} \quad (3.2)$$

where, i, j, and t denote city, industry, and year, respectively; empl represents the employment; robot indicates the number of industrial robots.

We evaluate urban exposure through both newly installed robots and the existing stock of robots, labeled as AI1 and AI2, respectively.

The Control Variables

Referring to Li et al. (2024) and Zeng and Zhang (2024), we incorporate various city-level variables to address potential confounding influences on the nexus between urban EE and AI. This chapter emphasizes the identification of significant and prevalent variables that influence urban EE through the lenses of economic, social, and demographic factors. Specifically, these variables encompass urban per capita GDP, the proportion of secondary industry in GDP, the ratio of fiscal revenue to GDP, the ratio of government spending on education to GDP, the ratio of government spending on technology to GDP, the number of college students per 10,000 urban residents, and the total volume of freight transported via highways.

Table 3.1 presents the basic statistical information for these variables. The average urban EE is relatively low, with a mean value of 0.538. Notably, among the 3948 observations of urban EE, only 31 demonstrate values equal to or greater than 1, indicating that these cities are on the DEA frontier and exhibit relatively high energy efficiency. This suggests

Table 3.1 Descriptive statistics

Variables	*Obs*	*Mean*	*S.D.*	*Min*	*Max*
	(1)	*(2)*	*(3)*	*(4)*	*(5)*
Core variables					
EE	3948	0.538	0.156	0.041	2.510
AI_1 (units/hundred workers)	3948	0.044	0.049	0.002	0.285
AI_2 (units/hundred workers)	3948	0.179	0.214	0.006	1.424
Control variables					
Urban per capita GDP (in log, RMB per capita)	3948	10.417	0.709	8.704	11.969
Proportion of the secondary industry in outputs	3948	0.479	0.109	0.117	0.910
Ratio of fiscal revenue to GDP	3948	0.071	0.027	0.021	0.240
Government spending on education as a share of GDP	3948	0.032	0.017	0.003	0.152
Government spending on technology as a share of GDP	3948	0.002	0.002	0.000	0.063
Number of college students per 10,000 urban residents (in log)	3948	12.951	1.205	9.812	15.556
Aggregate volume of freight transported on highways (in log, ten thousand tons)	3948	8.819	0.884	6.713	10.810

that the majority of Chinese cities experience low energy efficiency levels. Furthermore, for AI, the average density of installed industrial robots (AI_1) is 0.044 per 100 workers, whereas the stock of these robots (AI_2) stands at 0.179 per 100 workers. These results are consistent with the findings of Wei et al. (2020) and Zeng and Zhang (2024).

3.3.3 *Method*

We construct the following model to examine the influence of AI on urban EE:

$$\text{UEE}_{it} = \alpha_0 + \beta AI_{it} + \eta C_{it} + u_i + v_t + \varepsilon_{it} \tag{3.3}$$

where EE_{it} represents urban EE of city *i* in year *t*; AI_{it} indicates AI, quantified through both AI_1 and AI_2; β of interest captures the effect of AI on EE; *C* represents control variables; α_0 represents the intercepted item; u_i and v_t denote city and year fixed-effects (FE); ε_{it} is random error vector.

3.4 Findings

3.4.1 Baseline Results

We report the estimated results in Table 3.2. Columns (1) and (2) exclude control variables, while the remaining columns include all relevant variables. Fixed effects for both city and year are incorporated across all columns. The findings reveal that all AI coefficients are positive and statistically significant at the 1% level, demonstrating that AI can significantly enhance urban EE in Chinese cities, thereby supporting Hypothesis 3.1. Notably, the coefficient of AI_1 in Column (3) is 0.349, suggesting that when Chinese cities install one additional unit of industrial robots per hundred workers, their urban EE will increase by 0.349, which is equivalent to 64.87% of the average value (0.538). In other words, the installation of one additional standard deviation of industrial robots per hundred workers is associated with a 3.18% (0.049·64.87%) increase in their urban EE. Similarly, the coefficient of AI_2 in Column (4) is 0.083, indicating that each additional standard deviation of stock industrial robots per hundred workers can enhance urban EE by 3.30%.

The findings presented are consistent with existing scholarly literature (Dong et al., 2025). For example, Zeng and Wang (2025) find that AI can improve the energy efficiency of cities by promoting technological innovation and optimizing the industrial structure. Fan et al. (2025) identify the energy efficiency improvement effects of AI by leveraging the supply chain pilot policy as a quasi-natural experiment. Additionally, Hossin et al. (2023) project that energy efficiency in North Africa is anticipated to rise by 3.16% with a one standard deviation increase in the level of AI. Concurrently, Wang et al. (2022) highlight the critical contribution of AI to energy efficiency

Table 3.2 Baseline regression results

	(1)	*(2)*	*(3)*	*(4)*
AI_1	0.295***		0.349***	
	(3.566)		(4.129)	
AI_2		0.070***		0.083***
		(3.970)		(4.642)
Controls			√	√
City fixed-effect	√	√	√	√
Year fixed-effect	√	√	√	√
N	3948	3948	3948	3948
Adj. R^2	0.178	0.178	0.201	0.202

***$p < 0.01$. t-values are in parentheses

improvements, as evidenced by a single-factor analysis. Moreover, a multitude of studies has consistently illustrated that advanced technologies, including AI, provide significant benefits in mitigating the carbon intensity associated with energy consumption (Zhang & Zeng, 2024). Collectively, these findings substantiate the assertion that AI can effectively enhance urban EE.

3.4.2 Robustness Tests

Alternative Measure for Urban EE

In this chapter, the input-oriented radial distance function is employed in conjunction with the DEA measure of urban EE. This methodology is based on the premise that the DEA's objective is to minimize input usage while achieving predetermined output levels, with input variables adjusting in a proportional manner. However, this assumption may be overly stringent and inconsistent with actual conditions. Consequently, we have reassessed urban EE by employing a nonoriented and nonradial distance function, and further investigated the correlation between AI and urban EE. The results of this analysis are presented in Columns (1) and (2) of Table 3.3. Importantly, the coefficients for both AI_1 and AI_2 demonstrate significant positive values at conventional significance levels. These results support prior findings, thereby reinforcing the assertion that AI plays a role in improving urban EE in Chinese cities.

Table 3.3 Robustness tests I

	Alternative measure for urban EE		*Alternative measure for AI*	*Samples without the four municipalities*	
	(1)	*(2)*	*(3)*	*(4)*	*(5)*
AI_1	0.220*** (3.457)			0.327*** (3.865)	
AI_2		0.050*** (3.669)			0.079*** (4.383)
AI enterprise density			0.145*** (9.320)		
Controls	√	√	√	√	√
City fixed-effect	√	√	√	√	√
Year fixed-effect	√	√	√	√	√
N	3948	3948	3948	3892	3892
Adj. R^2	0.245	0.245	0.216	0.194	0.195

***$p < 0.01$. t-values are in parentheses

Alternative Measure for AI

This chapter reevaluates AI by measuring AI company density, defined as the number of AI companies per thousand urban residents. Data on these companies are gathered through a thorough search on the Tianyancha website.[5] The results are detailed in Column (3) of Table 3.3. Consistent with previous findings, these results suggest that AI contributes to enhancing urban EE in China.

Alternative Samples Without the Four Municipalities

The four municipalities—Beijing, Shanghai, Tianjin, and Chongqing—have more policy and resource support in China, which may contribute to the improvement of urban EE and AI. Consequently, the initial findings based on the samples covering these municipalities may overstate the influence of AI on urban EE. To mitigate this concern, we exclude these municipalities and reevaluate this effect. The findings from this reanalysis are detailed in Columns (4) and (5) of Table 3.3. Although the coefficients for AI_1 and AI_2 are slightly lower than those observed in the baseline results, they continue to be significantly positive and robust.

Panel Tobit Model

The positive values of urban EE may introduce biases when using OLS regression to assess the influence of AI on urban EE. Therefore, we adopt the Tobit method, which is a typical model for dealing with variables with limited values, to further investigate the influence of AI on urban EE. The results detailed in Columns (1) and (2) of Table 3.4 confirm the consistency of these findings with the baseline results, thereby reinforcing the robustness of our conclusions.

[5] Tianyancha: https://www.tianyancha.com/

(continued)

Table 3.4 Robustness tests II

	Panel Tobit model		DDL regression		IV method			
	(1)	(2)	(3)	(4)	Installation		Stock	
					(5)	(6)	(7)	(8)
					Stage 1: AI_1	Stage 2: EE	Stage 1: AI_2	Stage 2: EE
AI_1	0.292***		0.361***			0.299***		
	(3.531)		(4.155)			(2.604)		
AI_2		0.075***		0.087***				0.071***
		(4.242)		(4.716)				(2.707)
AI_{1iv}					0.238***			
					(56.116)			
AI_{2iv}							1.161***	
							(60.293)	
Controls	√	√	√	√	√	√	√	√
City fixed-effect	√	√	√	√	√	√	√	√
Year fixed-effect	√	√	√	√	√	√	√	√
N	3948	3948	3948	3948	3948	3948	3948	3948
Adj. R^2						0.201		0.202
Kleibergen-Paap rk LM statistic					475.890***		406.433***	
Cragg-Donald Wald F statistic					16000***		16000***	
Kleibergen-Paap rk Wald F statistic					3148.935***		3635.255***	

*** $p < 0.01$. t-values are in parentheses

DDL Regression

While we incorporate significant control variables at the city level, there remains a possibility that certain critical variables are absent, potentially resulting in bias in the regression coefficients of the core explanatory variables. We utilize the Double Debiased Lasso (DDL) method to address bias resulting from unobserved variables, accomplished through decomposing the singular values of regression matrix (Guo et al., 2022). The coefficients for AI1 and AI2 in

(continued)

(continued)

Columns (3) and (4) of Table 3.4 are marginally larger than those observed in the baseline regression. These findings indicate that there may be no significant omitted variable bias, thereby enhancing the reliability of our conclusions.

IV Method

In theory, urban EE has the potential to elevate the development of AI. More specifically, urban EE can significantly improve labor productivity and the competitive edge of cities. This improvement in urban comprehensive strength is likely to attract technological innovations, capital investments, and skilled professionals, thereby promoting the rapid advancement of AI capabilities within these cities. Should this hypothesis be validated, it may result in reversed causality, which could introduce bias into the regression outcomes. To address this potential concern, we employ the Instrumental Variable (IV) method.

Referring to Wang and Dong (2020), we leverage the US exposure to industrial robots at the industry level as the IVs. That is, we have AI_{1iv} and AI_{2iv} as the IVs of AI_1 and AI_2, respectively. The US exposure to industrial robots is closely related to that of China, as the United States is a major supplier of industrial robots imported by China. Importantly, the US exposure does not directly affect Chinese urban EE, making these IVs appropriate for our analysis. Additionally, we conduct under-identification and weak IV tests, with results presented in Table 3.4, confirming that the IVs do not exhibit signs of under-identification or weak instrument issues.

The results from the first stage demonstrate a significant positive correlation between IVs and both AI_1 and AI_2, aligning with our expectations. In the second stage, the coefficients for AI are all significantly positive, providing further evidence that AI can substantially improve urban EE. Thus, these findings reinforce our previous conclusions.

3.4.3 Heterogeneity Analysis

Region Heterogeneity
Economic conditions and technological capabilities exhibit considerable variation across different regions of China, with the eastern-central cities demonstrating a marked advantage over the western cities. The eastern-central cities, distinguished by their more advanced economic and technological landscapes, attract greater capital investment and promote technological innovation. As a result, these cities are more proficient in harnessing the potential of AI to improve urban EE in comparison to their western counterparts. In alignment with the classification established by the National Bureau of Statistics (NBS) of China, we categorize cities into two distinct groups: the eastern-central cities and the western cities.

The findings in Columns (1)–(2) of Table 3.5 suggest that the coefficients of AI_1 and AI_2 for the eastern-central regions are remarkably positive at 1% level, whereas those for western cities are negative and insignificant at the traditional level. This indicates that the influence of AI on urban EE is primarily evident in the eastern-central cities, while its effect in the western cities is minimal. The possible reason is that eastern-central cities benefit from superior foundational conditions, including advanced technology, state-of-the-art equipment, and strong policy support. These elements facilitate the development of AI technologies, thereby enhancing urban EE effectively. In contrast, the western cities face challenges such as a scarcity of AI expertise, limited funding, inadequate technological infrastructure, and constrained markets, which have significantly hindered the adoption of AI technologies. As a result, AI has not substantially impacted the urban EE in the western cities.

Table 3.5 Heterogeneity analysis results

	Region heterogeneity		*Location heterogeneity*		*Resource heterogeneity*		*Scale heterogeneity*	
	Eastern and Central	*Western*	*Northern*	*Southern*	*Resource-dependent*	*Non–resource-dependent*	*Type I: ≥5 million*	*Type II: <5 million*
	(1)	*(2)*	*(3)*	*(4)*	*(5)*	*(6)*	*(7)*	*(8)*
AI_1	0.541***	−0.041	0.825***	0.008	0.399***	0.118	0.681***	0.137
	(4.334)	(−0.284)	(6.998)	(0.061)	(2.621)	(1.050)	(4.612)	(1.304)
Adj. R^2	0.207	0.226	0.236	0.208	0.148	0.240	0.336	0.160
AI_2	0.119***	−0.002	0.183***	0.030	0.082**	0.046**	0.148***	0.041*
	(4.463)	(−0.056)	(7.210)	(1.117)	(2.506)	(1.977)	(4.798)	(1.825)
Adj. R^2	0.207	0.226	0.237	0.208	0.148	0.241	0.337	0.161
Controls	√	√	√	√	√	√	√	√
City fixed-effect	√	√	√	√	√	√	√	√
Year fixed-effect	√	√	√	√	√	√	√	√
N	2548	1400	2030	1918	1596	2352	1414	2534

***$p < 0.01$. **$p < 0.05$. *$p < 0.1$. t-values are in parentheses

Location Heterogeneity
Significant disparities exist in the natural conditions between cities situated in southern and northern China, leading to distinct industrial structures and lifestyles. For example, the prevalence of coal resources in northern cities has facilitated the development of heavy industries. Additionally, the low winter temperatures in these regions necessitate the installation of heating systems to mitigate the cold, which can result in considerable emissions of pollutants. Conversely, southern cities typically emphasize light industry and exhibit a reduced requirement for winter heating. These contrasting characteristics contribute to a comparatively lower urban EE in northern cities when juxtaposed with their southern counterparts. As a result, the impact of AI on urban EE may vary across these cities. Therefore, we classify the cities in our sample into northern and southern groups, delineated by the geographical boundary established by the Qinling Mountains–Huaihe River Line in China.

We report the results in Columns (3) and (4) of Table 3.5. The results indicate that the coefficients for AI_1 and AI_2 in northern cities are significant and positive at conventional levels, whereas those in southern cities do not reach significance. This suggests that AI has substantially improved urban EE in northern cities, while its effect in southern cities is minimal. Several underlying factors may explain this disparity. Initially, the comparatively low urban EE observed in northern cities, which can be attributed to their industrial composition and lifestyle choices, presents considerable opportunities for the integration of AI to improve urban EE. Conversely, southern cities are predominantly defined by light industries and a multitude of labor-intensive sectors, where the extensive implementation of AI is anticipated to supplant human labor. This energy-centric characteristic of AI engenders a substitution effect that hinders advancements in urban

Resource Heterogeneity
Resource-dependent cities are defined by their dependence on industries focused on the extraction and processing of natural resources. Significant differences exist between resource-dependent and non–resource-dependent cities in terms of energy resources,

(continued)

(continued)

industrial composition, and economic development pathways, which may affect the influence of AI on urban EE. In accordance with the National Sustainable Development Plan of China, we classify the samples into two categories: resource-dependent cities and non–resource-dependent cities.

EE. Consequently, there is a lack of definitive evidence supporting the notion that AI can enhance urban EE in southern cities.

We present the estimates in Columns (5) and (6) of Table 3.5. In resource-dependent cities, both AI_1 and AI_2 exhibit significant positive coefficients, indicating that AI substantially enhances urban EE in these areas. On the contrary, the coefficients for non–resource-dependent cities show inconsistency; specifically, AI_2 is significant and positive at the 5% level, while AI_1 lacks statistical significance. This suggests that the influence of AI on non–resource-dependent cities is not robust. Additionally, the coefficients for both AI_1 and AI_2 in resource-dependent cities surpass those found in non–resource-dependent cities. This indicates that the influence of AI on urban EE is primarily evident in resource-dependent cities, which are marked by high energy consumption, elevated pollution levels, and increased emissions. Our results align with the findings of Li et al. (2024), which indicate that AI is more advantageous for enhancing energy efficiency in resource-dependent cities compared to non–resource-dependent cities. In fact, the implementation of AI technology can effectively drive technological progress and industrial transformation in these cities, aiding

Scale Heterogeneity

The scale of a city significantly determines the extent to which its economic scale effects can be realized. Larger cities are generally better positioned to achieve efficient resource allocation and capitalize on economies of scale. Consequently, the impact of AI on urban EE may differ based on city size. Referring to the 2014 Notice regarding the classification of city scales, samples are categorized into five types: small cities (fewer than 0.5 million), medium-sized cities (0.5 to less than 1 million), large cities (1 to less than 5 million), mega

(continued)

(continued)

cities (5 to less than 10 million), and super cities (10 million or more). Considering the limited sample size, we combine small and medium-sized cities into large city type and combine super cities into mega cities. Finally, this chapter identifies two categories: Type I cities (five million or more) and Type II cities (fewer than five million).

their transition toward a green economy and sustainable energy practices. Thus, AI is essential for enhancing urban EE in resource-dependent cities.

We report the results in Table 3.5. The findings in Columns (7) and (8) indicate that Type I cities exhibit greater robustness than Type II cities, characterized by larger coefficients and higher levels of statistical significance. That is, the urban EE of Type I cities can be significantly improved by AI, while its impact on Type II cities remains uncertain. The observed discrepancy can be attributed to two primary factors. Firstly, Type I cities typically have access to advanced technology, substantial financial resources, and an abundance of assets. These factors contribute to the advancement and innovation of AI technologies, thereby enhancing urban EE more effectively. Secondly, Type I cities are characterized by a well-developed industrial ecosystem, which promotes the integration of various industries and allows for the realization of synergistic effects that capitalize on economies of scale. As a result, this synergy can effectively leverage the capabilities of AI, leading to a significant improvement in urban EE.

3.4.4 Estimates for Spatial Spillover Effects

Spatial Autocorrelation Analysis

The examination of spatial effects necessitates the presence of spatial autocorrelation among the primary variables, signifying that local cities and their adjacent counterparts display correlated values. Prior to conducting spatial autocorrelation assessments, it is imperative to develop spatial weight matrices that encapsulate the complex interrelations among cities. These matrices serve to quantify the proximity between cities, thereby elucidating the concept of "neighboring." Drawing on the works of Fingleton and Gallo (2008) and Zhang

(continued)

(continued)

et al. (2023b), we construct two types of spatial weight matrices: spatial contiguity weight matrix $\mathbf{W_1}$ and spatial nested weight matrix $\mathbf{W_2}$. We set the matrices as follows:

$$\mathbf{W}_1 : w_{ij} = \begin{cases} 1, i \text{ is adjacent to } j, \text{and } i \neq j \\ 0, \text{else} \end{cases}$$
$$\mathbf{W}_2 : w_{ij} = \begin{cases} \dfrac{1}{\left|\text{GDP}_i - \text{GDP}_j - 1\right| \times e^{d_{ij}}}, i \neq j \\ 0, i = j \end{cases} \tag{3.4}$$

Where i and j denote city; GDP represents the average GDP from 2006 to 2019; d_{ij} indicates the geographical distance.

Table 3.6 Global Moran's Index results

	W_1			W_2		
	EE	AI_1	AI_2	*EE*	AI_1	AI_2
	(1)	(2)	(3)	(4)	(5)	(6)
2006	0.079**	0.393***	0.224***	0.009	0.202***	0.098***
2007	0.087**	0.390***	0.274***	−0.008	0.160***	0.034
2008	0.107***	0.383***	0.311***	0.007	0.130***	0.036
2009	0.153***	0.350***	0.318***	0.053**	0.065***	0.038
2010	0.167***	0.384***	0.383***	−0.004	0.257***	0.133***
2011	0.214***	0.385***	0.392***	0.043*	0.254***	0.192***
2012	0.209***	0.385***	0.392***	0.058**	0.255***	0.214***
2013	0.146***	0.387***	0.391***	0.015	0.246***	0.226***
2014	0.164***	0.383***	0.389***	0.034	0.261***	0.240***
2015	0.214***	0.383***	0.387***	0.068**	0.260***	0.247***
2016	0.129***	0.387***	0.386***	0.019	0.245***	0.248***
2017	0.178***	0.387***	0.386***	0.078***	0.247***	0.249***
2018	0.236***	0.387***	0.386***	0.149***	0.242***	0.248***
2019	0.293***	0.387***	0.386***	0.127***	0.238***	0.246***

***$p < 0.01$. **$p < 0.05$. *$p < 0.1$

This chapter performs spatial autocorrelation tests on the core variables utilizing the Global Moran's Index. We report the results in Table 3.6. T The Moran's Index values are predominantly positive, and most are significant at 1% level. These results suggest a positive spatial autocorrelation between urban EE and AI in Chinese cities, characterized by clusters of high and low values. Consequently, the application of spatial analysis in this research is both justified and essential.

This chapter leverages local Moran's index scatterplots to describe the agglomeration characteristics of the sample cities. The first and third quadrants of these plots reveal that the correlation between adjacent cities exhibits a positive trend, while the others reveal a negative trend. Figure 3.2 illustrates the Moran's index scatterplots for core variables in 2019. Most cities are positioned in either the first or the third quadrant, indicating a clear pattern of low-value cluster pairing, alongside high-value cluster coexistence. These findings emphasize the significance of spatial analysis in assessing the impact of AI on urban EE.

Spatial Econometric Model

Commonly used spatial econometric models are the Spatial Autoregressive Model (SAR), the Spatial Error Model (SEM), and the SDM. To identify the most appropriate model from these alternatives, we utilize Likelihood Ratio (LR) tests and Wald tests.

These significant values of LR_lag and Wald_lag in Table 3.7 mean that SDM cannot degradation into SAR. Similarly, the results of LR_error and Wald_error prove that SDM cannot degradation into SEM. Therefore, these findings evidence that this chapter is better suited for constructing SDM to investigate the spatial spillover effects of AI on urban EE, rather than employing SAR or SEM.

We construct SDM as follows:

$$\begin{aligned} \mathrm{UEE}_{it} &= \alpha_0 + \rho W \times \mathrm{UEE}_{it} + \beta AI_{it} + \beta_1 W \times AI_{it} + \eta C_{it} \\ &+ \eta_1 W \times C_{it} + u_i + v_t + \varepsilon_{it} \end{aligned} \tag{3.5}$$

where β_1 captures the spatial spillover effects; **W** indicates the spatial weight matrix.

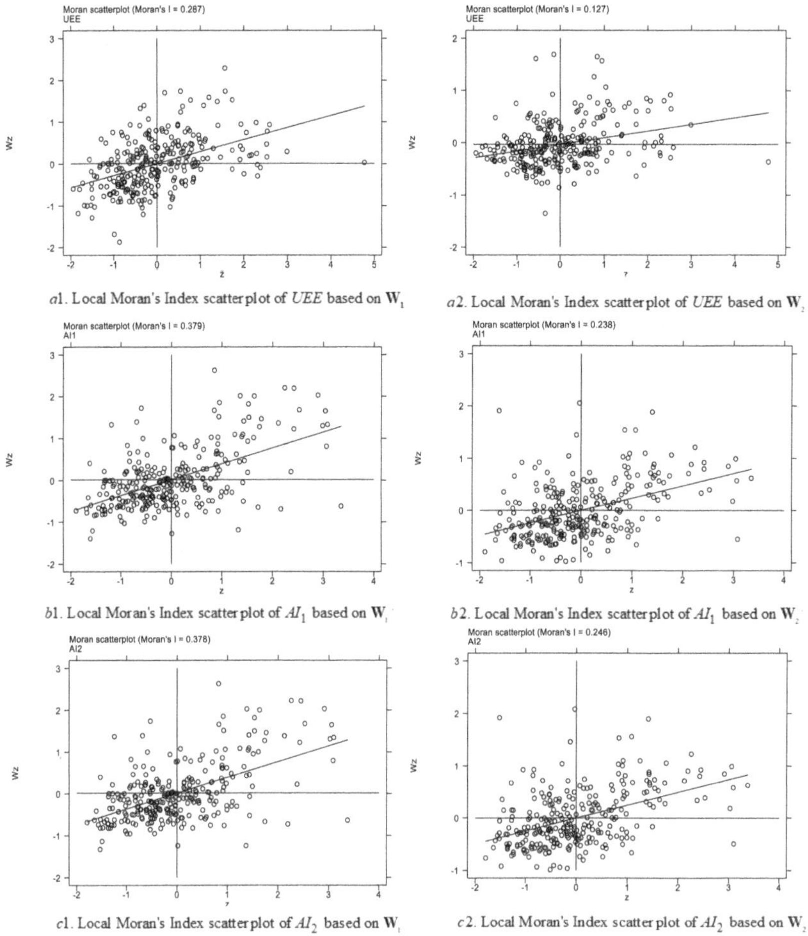

Fig. 3.2 Local Moran's index in 2019

Table 3.7 Results of LR tests and Wald tests

	W_1		W_2	
	AI_1	AI_2	AI_1	AI_2
	(1)	(2)	(3)	(4)
LR_lag	152.46***	154.93***	151.99***	153.95***
LR_error	148.28***	150.77***	150.83***	153.30***
Wald_lag	155.38***	158.14***	154.95***	156.97***
Wald_error	143.82***	146.86***	152.43***	154.85***

***$p < 0.01$

Spatial Analysis Results

We report the estimates in Table 3.8, where Columns (1)–(2) correspond to **W1** and Columns (3)–(4) pertain to $\mathbf{W_2}$. The coefficients for AI_1 and AI_2, representing the impact of local AI on urban EE, are positive, aligning with earlier findings. The coefficients for $AI_1 \times \mathbf{W_1}$, $AI_2 \times \mathbf{W_1}$, $AI_1 \times \mathbf{W_2}$, and $AI_2 \times \mathbf{W_2}$ capture the spatial spillover effects, all of which are significant and positive at the 5% level. These results indicate that AI generates beneficial spatial spillover effects on urban EE, confirming Hypothesis 3.2. Specifically, the urban EE of local cities can be enhanced by the AI systems of neighboring cities, with a more pronounced effect observed in areas where neighboring cities exhibit higher levels of AI sophistication and utilization. For instance, the coefficient of $AI_1 \times \mathbf{W_1}$ in Column (1) is 0.263, indicating that for each additional unit of industrial robots installed per hundred workers in neighboring cities, the energy efficiency of local cities increases by 0.263. The remaining coefficients can be interpreted similarly.

(continued)

(continued)

Table 3.8 Results of SDM regression

	W_1		W_2	
	(1)	*(2)*	*(3)*	*(4)*
AI_1	0.346***		0.125	
	(3.687)		(1.418)	
AI_2		0.080***		0.026
		(4.021)		(1.373)
$AI_1 \times W_1$	0.263**			
	(2.151)			
$AI_2 \times W_1$		0.062**		
		(2.366)		
$AI_1 \times W_2$			0.384**	
			(2.200)	
$AI_2 \times W_2$				0.112***
				(3.039)
rho	0.089***	0.089***	0.080***	0.076***
	(3.751)	(3.735)	(3.160)	(2.998)
sigma2_e	0.006***	0.006***	0.006***	0.006***
	(44.399)	(44.394)	(44.413)	(44.414)
Controls	√	√	√	√
City fixed-effect	√	√	√	√
Year fixed-effect	√	√	√	√
N	3948	3948	3948	3948
R^2	0.006	0.006	0.061	0.056

***$p < 0.01$. **$p < 0.05$. t-values are in parentheses

Meanwhile, we assess the robustness of the aforementioned conclusions by employing the method presented by Sage and Pace (2009). In fact, deviations may arise when employing SDM regression coefficients to capture the spatial spillover effects due to the complex interactions between local cities and their neighbors. To avoid these possible deviations, Sage and Pace (2009) proposed the spatial effect decomposition of SDM to accurately capture its spatial spillover effect. Table 3.9 reports the decomposition results, which include direct effects measuring how AI in local cities affects their own urban EE, indirect effects assessing how AI in neighboring cities impacts local urban EE, and total effects reflecting the

Table 3.9 Results of spatial effect decomposition of SDM

	W_1		W_2	
	AI_1	AI_2	AI_1	AI_2
	(1)	(2)	(3)	(4)
Direct effects	0.356***	0.083***	0.134	0.028
	(3.726)	(4.061)	(1.483)	(1.454)
Indirect effects	0.319***	0.075***	0.440**	0.125***
	(2.584)	(2.829)	(2.434)	(3.302)
Total effects	0.675***	0.157***	0.574***	0.153***
	(5.718)	(6.259)	(3.223)	(4.184)

***$p < 0.01$. **$p < 0.05$. t-values are in parentheses

overall impact of AI across all cities on local urban EE. Notably, the indirect effects are significant at the 5% level, supporting the conclusion that AI generates positive spatial spillover effects on urban EE.

Our findings align with existing research, providing mutual reinforcement. Relevant studies indicate that AI has spatial spillover effects in reducing carbon emission intensity. For example, a comparable finding was reported by Li et al. (2024), which indicated that the impact of AI on urban EE exhibits a substantial positive spatial spillover effect. Specifically, the local level of AI contributes to the enhancement of energy efficiency in adjacent cities. Ding et al. (2023) validate the potential of AI to drive reductions in carbon emission intensity across neighboring provinces, supported by provincial data from China. Similarly, Tao et al. (2023) reveal that AI effectively mitigates carbon emissions not only from local industries but also from adjacent industrial sectors. Carbon emissions are a vital indicator for assessing energy efficiency, and a clear correlation exists between carbon emission intensity and energy efficiency. Consequently, local AI initiatives may improve urban EE in neighboring cities.

Several justifications support the assertion of spatial spillover effects of AI on urban EE. On the one hand, neighboring cities can benefit passively when local cities implement AI to enhance their urban EE (Zhang et al., 2023a). Neighboring cities frequently exhibit strong economic interconnections, typically functioning as components of each other's industrial supply chains. Consequently, enhancements in energy efficiency within one segment of the chain tend to result in corresponding improvements in energy efficiency in adjacent locations. This phenomenon explains why an increase in urban EE, driven by AI in a local city, tends to propagate to its neighboring cities. On the other hand, neighboring cities can actively improve their urban EE by adopting AI technologies from local cities and fostering economic collaboration. A phenomenon frequently observed among adjacent urban areas is the interplay of a leading and catching-up effect (Tao et al., 2023). Specifically, when a local city adopts advanced AI technologies to enhance its energy efficiency, neighboring cities tend to engage in active learning from this example and foster collaborative efforts to elevate their own energy efficiency levels.

3.5 Discussions

3.5.1 Contributions

This chapter offers several key contributions to this field. First, this chapter provides a new research perspective and enriches the relevant research results by exploring the AI-urban EE nexus. In recent years, the academic community has demonstrated considerable enthusiasm for research in AI, yielding impressive results. However, there remains a paucity of studies that thoroughly investigate the impact of AI on urban EE (Li et al., 2024). While some research has examined the effects of AI on energy intensity and carbon emissions within the manufacturing sector (Zhou et al., 2024), and other studies have explored the relationship between AI and energy efficiency across various levels—such as provinces (Li et al., 2025) and enterprises (Song et al., 2024)—to date, only one article has been found that directly addresses the connection between AI and urban EE (Li et al., 2024). They find that AI helps improve urban EE using data from Chinese cities; however, they fail to investigate the potential variability in the effects of AI on urban EE across diverse urban settings. In fact, a comprehensive analysis of AI's influence on energy efficiency at the city level is both necessary and significant for two primary reasons. First, given that AI has the potential to both promote energy conservation and increase energy consumption, its overall effect on urban EE remains ambiguous. Thus, investigating the influence of AI on urban EE is essential. Second, given that cities are major hubs of energy consumption, the efficiency of energy use within urban areas is intrinsically linked to the overall energy efficiency of the economy.

Second, this chapter presents a novel approach to research methodologies. Utilizing empirical data from Chinese cities, this chapter employs econometric techniques to investigate the influence of AI on urban EE. While existing literature predominantly focuses on theoretical frameworks for the development of AI and its enhancement of urban EE—addressing factors such as the institutional environment, technological infrastructure, and human capital—there is a notable scarcity of empirical studies examining the relationship between AI and urban EE. This research not only implements a classical fixed effects model for benchmark regression analysis but also conducts a comprehensive robustness assessment through various methodologies. Especially, to mitigate potential regression bias stemming from omitted variable issues, this chapter incorporates an advanced DDL model proposed by Guo et al. (2022) for robustness testing, thereby bolstering the validity of its findings. Furthermore, this chapter employs spatial econometric techniques to investigate the spatial spillover effects of AI on urban EE.

Third, there are some innovations in the mechanism of this chapter. This chapter innovatively analyzes the heterogeneous influences across different types of cities. The findings indicate that resource-dependent cities, as well as cities located in eastern-central regions, northern cities of the Qinling Mountains–Huaihe River Line, and both super-large and large cities experience more pronounced enhancements in e-commerce and urban EE following the adoption of AI technology. This insight provides a scientific basis for developing differentiated urban energy policies and highlights the significance of promoting AI technologies that are tailored to specific urban characteristics. Furthermore, this chapter systematically examines the spatial spillover effects. These findings elucidate the interconnectedness of AI technology across urban areas, offering a new perspective on its role in regional economic development and providing strategic recommendations for fostering regional collaborative growth.

3.5.2 Limitations

This chapter examines the influence of AI on urban EE at the city level and presents certain findings. Nonetheless, this chapter is subject to several limitations, primarily in the following areas: as an empirical investigation grounded in observational data, it does not provide a comprehensive analysis of viable strategies for enhancing both the AI capabilities and energy efficiency within Chinese cities. The findings contribute to the understanding of how urban AI can be leveraged to improve urban EE; therefore, future research should delve deeper into methods for advancing urban AI. This includes, but is not limited to, enhancing AI-related infrastructure, fostering the development of AI talent, strengthening data privacy protection systems associated with AI, and improving the readiness of urban energy systems to integrate AI technologies. These topics represent potential avenues for further exploration and development stemming from this chapter.

3.5.3 Policy Implications

First, the government should implement incentive policies, such as tax incentives, financial subsidies, and patent protection, to facilitate the extensive and profound application of AI in Chinese cities. This can enhance the green production technology of enterprises and revolutionize urban residents' lifestyles, facilitate the realization of environmentally sustainable industrial development, and accelerate the eco-friendly transformation of the economy, thereby effectively enhancing urban EE. The proposed measures encompass several key components. Firstly, there is a need to expedite the enhancement of the city's technical infrastructure, with a particular emphasis on advanced digital technology facilities, while also fostering the integration and application of AI technologies in the operations of enterprises and the daily lives of residents. Secondly, it is essential to establish and refine systems for the protection of AI-related data, as well as frameworks for patent protection, laws, and regulations. These institutional arrangements are critical in providing the necessary foundation and impetus for the city to elevate its AI capabilities. Lastly, it is imperative to bolster the preparedness of the existing energy systems to incorporate AI technologies. This entails a comprehensive upgrade and reform of China's current energy infrastructure to enhance the utilization of AI in various domains, including energy management, monitoring, transportation, and forecasting.

Second, the consideration of city heterogeneity is crucial, as it necessitates the development of appropriate strategies and policies that align with the unique circumstances of various kinds of cities. The government should allocate special attention to western cities, southern cities, non–resource-dependent cities, and smaller cities due to the limited impact of AI on their energy efficiency. In the context of China's western cities, it is imperative for the government to enhance financial investments in AI-related infrastructure. This should encompass the introduction of advanced technologies, fostering innovation, and facilitating talent development and technical exchanges. Notably, public resources should be preferentially allocated to these western cities. Conversely, for southern cities and non–resource-dependent cities in China, it is crucial to address the potential increase in energy consumption resulting from the advancement of AI technologies. Preparations must be made for the energy supply systems, which include securing energy raw materials, ensuring the availability of skilled practitioners, and enhancing management capabilities. Additionally, timely measures should be implemented to mitigate the environmental impacts associated with heightened energy consumption. Furthermore, small cities in China should accelerate their industrial development, encourage industrial clustering, and cultivate economies of scale and industrial synergies to maximize the benefits derived from AI utilization.

Third, the government can fully leverage the spatial spillover effect of AI to enhance urban EE. The government can prioritize the development of AI in selected key cities and facilitate its diffusion to neighboring cities, thereby promoting energy efficiency enhancement across all cities. The government can initiate pilot projects in select major cities, implement demonstration projects for AI development, and promote engagement from both enterprises and residents. Concurrently, enterprises are encouraged to engage in comprehensive collaboration with universities and research institutions to collectively advance the research, development, and implementation of intelligent technologies. Furthermore, it is advisable to enhance the impact of these pilot cities, enabling them to spearhead the modernization of the energy systems in adjacent cities, thereby improving urban energy efficiency in the broader regions.

3.6 Conclusions

Energy consumption in China is primarily concentrated in urban regions; however, the urban EE remains considerably lower than the global average. Recent developments in AI present a significant opportunity to improve urban EE. Nonetheless, the impact of AI on urban EE is ambiguous due to its dual role as both a promoter of energy conservation and a substantial energy consumer. In light of this, this chapter utilizes data from 282 Chinese cities covering the period from 2006 to 2019 to investigate the influence of AI on urban EE. The results indicate that AI contributes positively to the enhancement of urban EE in China. Specifically, the introduction of one additional standard deviation of industrial robots per hundred workers correlates with a 3.18% increase in urban EE. Furthermore, each additional standard deviation of stock industrial robots per hundred workers is associated with a 3.30% improvement in urban EE. These findings are robust across various measurement techniques, including the exclusion of municipal samples and the application of alternative methodologies such as the panel Tobit model, DDL regression, and IV methods. Additionally, this chapter demonstrates that the positive impact of AI on urban EE is further amplified in eastern-central cities, northern cities, resource-dependent cities, as well as mega and super-sized cities. Finally, the research establishes that AI generates positive spatial spillover effects on urban EE, whereby the energy efficiency of local cities can be enhanced through the influence of AI systems in neighboring cities, with a more pronounced effect observed when these neighboring cities exhibit higher levels of AI sophistication and widespread implementation.

References

Acemoglu, D., & Restrepo, P. (2020). Robots and jobs: Evidence from US labor markets. *Journal of Political Economy, 128*(6), 2188–2244. https://doi.org/10.1086/705716

Ahmad, T., Zhang, D. D., Huang, C., Zhang, H. C., Dai, N. Y., Song, Y. H., & Chen, H. X. (2021). Artificial intelligence in sustainable energy industry: Status Quo, challenges and opportunities. *Journal of Cleaner Production, 289*, Article 125834. https://doi.org/10.1016/j.jclepro.2021.125834

Allam, Z., & Dhunny, Z. A. (2019). On big data, artificial intelligence and smart cities. *Cities, 89*, 80–91. https://doi.org/10.1016/j.cities.2019.01.032

Anthopoulos, L., & Kazantzi, V. (2022). Urban energy efficiency assessment models from an AI and big data perspective: Tools for policy makers. *Sustainable Cities and Society, 76*, Article 103492. https://doi.org/10.1016/j.scs.2021.103492

Benzidia, S., Makaoui, N., & Bentahar, O. (2021). The impact of big data analytics and artificial intelligence on green supply chain process integration and hospital environmental performance. *Technological Forecasting and Social Change, 165*, Article 120557. https://doi.org/10.1016/j.techfore.2020.120557

Berg, A., Buffie, E. F., & Zanna, L. F. (2018). Should we fear the robot revolution? (the correct answer is yes). *Journal of Monetary Economics, 97*, 117–148. https://doi.org/10.1016/j.jmoneco.2018.05.014
Bibri, S. E., Huang, J. F., & Krogstie, J. (2024). Artificial intelligence of things for synergizing smarter eco-city brain, metabolism, and platform: Pioneering data-driven environmental governance. *Sustainable Cities and Society, 108*, Article 105516. https://doi.org/10.1016/j.scs.2024.105516
Boza, P., & Evgeniou, T. (2021). Artificial intelligence to support the integration of variable renewable energy sources to the power system. *Applied Energy, 290*, Article 116754. https://doi.org/10.1016/j.apenergy.2021.116754
Chang, L., Taghizadeh-Hesary, F., & Mohsin, M. (2023). Role of artificial intelligence on green economic development: Joint determinates of natural resources and green total factor productivity. *Resources Policy, 82*, Article 103508. https://doi.org/10.1016/j.resourpol.2023.103508
Chang, Q. Y., Wu, J., & Zhang, Y. (2024). Strengthening energy transition through effective urban electric demand prediction: A multidisciplinary exploration of AI, policy, and administrative strategies. *Sustainable Cities and Society, 107*, Article 105378. https://doi.org/10.1016/j.scs.2024.105378
Chen, W. Y. (2015). The role of urban green infrastructure in offsetting carbon emissions in 35 major Chinese cities: A nationwide estimate. *Cities, 44*, 112–120. https://doi.org/10.1016/j.cities.2015.01.005
Chen, Z., & Liu, Y. S. (2021). Economic development ideas of regional convergence industry based on artificial intelligence multimedia background. *Mobile Information Systems, 11*, Article 4787858. https://doi.org/10.1155/2021/4787858
Chen, L. F., & Wang, K. F. (2022). The spatial spillover effect of low-carbon city pilot scheme on green efficiency in China's cities: Evidence from a quasi-natural experiment. *Energy Economics, 110*, Article 106018. https://doi.org/10.1016/j.eneco.2022.106018
Chen, W. X., Wang, G. Z., Xu, N., Ji, M., & Zeng, J. (2024). Promoting or inhibiting? New-type urbanization and urban carbon emissions efficiency in China. *Cities, 140*, Article 104429. https://doi.org/10.1016/j.cities.2023.104429
Cui, H. Y., & Cao, Y. Q. (2024). Do smart cities improve energy efficiency? A test of spatial effects and mechanisms. *Sustainable Cities and Society, 101*, Article 105124. https://doi.org/10.1016/j.scs.2023.105124
Dauvergne, P. (2022). Is artificial intelligence greening global supply chains? Exposing the political economy of environmental costs. *Review of International Political Economy, 29*(3), 696–718. https://doi.org/10.1080/09692290.2020.1814381

Ding, T., Li, J. Y., Shi, X., Li, X. H., & Chen, Y. (2023). Is artificial intelligence associated with carbon emissions reduction? Case of China. *Resources Policy, 85*, Article 103892. https://doi.org/10.1016/j.resourpol.2023.103892

Dong, M. F., Wang, G., & Han, X. F. (2024). Impacts of artificial intelligence on carbon emissions in China: In terms of artificial intelligence type and regional differences. *Sustainable Cities and Society, 113*, Article 105682. https://doi.org/10.1016/j.scs.2024.105682

Dong, K., Yang, Y. Y., Sheng, T. X., Liu, L. Y., & Song, K. Y. (2025). How does artificial intelligence affect sustainable development by improving energy efficiency and optimizing energy structure. *Environment Development and Sustainability*. https://doi.org/10.1007/s10668-025-06618-1

Du, M., Zhang, Y. F., Dong, H., & Zhou, X. J. (2023). Heterogeneous impact of artificial intelligence on carbon emission intensity: Empirical test based on provincial panel data in China. *Frontiers in Ecology and Evolution, 11*, Article 1058505. https://doi.org/10.3389/fevo.2023.1058505

Fan, J. J., Li, W. Q., & Chen, L. (2025). How does artificial intelligence affect energy efficiency? Evidence from supply chain digitization pilot program. *Energy Economics, 149*, Article 108728. https://doi.org/10.1016/j.eneco.2025.108728

Fingleton, B., & Gallo, J. L. (2008). Estimating spatial models with endogenous variables, a spatial lag and spatially dependent disturbances: Finite sample properties. *Papers in Regional Science, 87*(3), 319–340. https://doi.org/10.1111/j.1435-5957.2008.00187.x

Gao, D., Li, Y., & Li, G. (2022a). Boosting the green total factor energy efficiency in urban China: Does low-carbon city policy matter? *Environmental Science and Pollution Research, 29*(37), 56341–56356. https://doi.org/10.1007/s11356-022-19553-9

Gao, D., Li, G., & Yu, J. Y. (2022b). Does digitization improve green total factor energy efficiency? Evidence from Chinese 213 cities. *Energy, 247*, Article 123395. https://doi.org/10.1016/j.energy.2022.123395

Guo, Z. J., Ćevid, D., & Bühlmann, P. (2022). Doubly Debiased LASSO: High-dimensional inference under hidden confounding. *The Annals of Statistics, 50*(3), Article 1320. https://doi.org/10.1214/21-aos2152

Hong, Q. Q., Cui, L. H., & Hong, P. H. (2022). The impact of carbon emissions trading on energy efficiency: Evidence from quasi-experiment in China's carbon emissions trading pilot. *Energy Economics, 110*, Article 106025. https://doi.org/10.1016/j.eneco.2022.106025

Hossin, M. A., Alemzero, D., Wang, R. P., Kamruzzaman, M. M., & Mhlanga, M. N. (2023). Examining artificial intelligence and energy efficiency in the MENA region: The dual approach of DEA and SFA. *Energy Reports, 9*, 4984–4994. https://doi.org/10.1016/j.egyr.2023.03.113

Jung, J. H., & Lim, D. G. (2020). Industrial robots, employment growth, and labor cost: A simultaneous equation analysis. *Technological Forecasting and Social Change, 159*, Article 120202. https://doi.org/10.1016/j.techfore.2020.120202

Khavarian-Garmsir, A. R., Sharifi, A., & Sadeghi, A. (2023). The 15-minute city: Urban planning and design efforts toward creating sustainable neighborhoods. *Cities, 132*, Article 104101. https://doi.org/10.1016/j.cities.2022.104101

Leal, W., Mbah, M. F., Dinis, M. A. P., Trevisan, L. V., de Lange, D., Mishra, A., Rebelatto, B., Ben-Hassen, T., & Aina, Y. A. (2024). The role of artificial intelligence in the implementation of the UN sustainable development goal 11: Fostering sustainable cities and communities. *Cities, 150*, Article 105021. https://doi.org/10.1016/j.cities.2024.105021

Lee, C. C., & Hussain, J. (2023). Energy sustainability under the COVID-19 outbreak: Electricity break-off policy to minimize electricity market crises. *Energy Economics, 125*, Article 106870. https://doi.org/10.1016/j.eneco.2023.106870

Lee, C. C., & Lee, C. C. (2022). How does green finance affect green total factor productivity? Evidence from China. *Energy Economics, 107*, Article 105863. https://doi.org/10.1016/j.eneco.2022.105863

Li, Q. Y., Zhang, J. Q., Feng, Y., Sun, R. G., & Hu, J. (2024). Towards a high-energy efficiency world: Assessing the impact of artificial intelligence on urban energy efficiency. *Journal of Cleaner Production, 461*, Article 142593. https://doi.org/10.1016/j.jclepro.2024.142593

Li, X., Li, S. Y., Cao, J. F., & Spulbar, A. C. (2025). Does artificial intelligence improve energy efficiency? Evidence from provincial data in China. *Energy Economics, 142*, Article 108149. https://doi.org/10.1016/j.eneco.2024.108149

Liu, L., Yang, K., Fujii, H., & Liu, J. (2021). Artificial intelligence and energy intensity in China's industrial sector: Effect and transmission channel. *Economic Analysis and Policy, 70*, 276–293. https://doi.org/10.1016/j.eap.2021.03.002

Liu, J., Liu, L., Qian, Y., & Song, S. F. (2022). The effect of artificial intelligence on carbon intensity: Evidence from China's industrial sector. *Socio-Economic Planning Sciences, 83*, Article 101002. https://doi.org/10.1016/j.seps.2020.101002

Luo, T., Wong, W. F., & Yau, W. Y. (2023). Achieving green AI with energy-efficient deep learning using neuromorphic computing. *Communications of the ACM, 66*(7), 52–57. https://doi.org/10.1145/3588591

Luo, K. K., Wang, F., & Chen, X. Z. (2025). Impact of artificial intelligence on energy efficiency in Chinese enterprises. *International Review of Economics & Finance, 103*, Article 104541. https://doi.org/10.1016/j.iref.2025.104541

Ma, R. Y., & Lin, B. Q. (2023). Digital infrastructure construction drives green economic transformation: Evidence from Chinese cities. *Humanities & Social Sciences Communications, 10*(1), Article 460. https://doi.org/10.1057/s41599-023-01839-z

Ma, L. D., Hong, Y. X., He, S. C., Luo, H. B., Liu, G. M., Zheng, J. H., Xia, Y. T., & Xiao, D. H. (2023). The influence of digital economy development on urban carbon emission intensity in the Yangtze River Economic Belt: Mediating mechanism and spatial effect. *Frontiers in Ecology and Evolution, 11*, Article 1148505. https://doi.org/10.3389/fevo.2023.1148505

Makridis, C. A., & Mishra, S. (2022). Artificial intelligence as a service, economic growth, and well-being. *Journal of Service Research, 25*(4), 505–520. https://doi.org/10.1177/10946705221120218

Maleki, R., Asadnia, M., & Razmjou, A. (2022). Artificial intelligence-based material discovery for clean energy future. *Advanced Intelligent Systems, 4*(10), Article 2200073. https://doi.org/10.1002/aisy.202200073

Mehmood, M. U., Chun, D., Zeeshan, Han, H., Jeon, G., & Chen, K. (2019). A review of the applications of artificial intelligence and big data to buildings for energy-efficiency and a comfortable indoor living environment. *Energy Buildings, 202*, Article 109383. https://doi.org/10.1016/j.enbuild.2019.109383

Ounifi, H. A., Gherbi, A., & Kara, N. (2022). Deep machine learning-based power usage effectiveness prediction for sustainable cloud infrastructures. *Sustainable Energy Technologies and Assessment, 52*, Article 101967. https://doi.org/10.1016/j.seta.2022.101967

Pan, S. L., & Nishant, R. (2023). Artificial intelligence for digital sustainability: An insight into domain-specific research and future directions. *International Journal of Information Management, 72*, Article 102668. https://doi.org/10.1016/j.ijinfomgt.2023.102668

Qu, C. Y., Shao, J., & Shi, Z. K. (2020). Does financial agglomeration promote the increase of energy efficiency in China? *Energy Policy, 146*, Article 111810. https://doi.org/10.1016/j.enpol.2020.111810

Regona, M., Yigitcanlar, T., Hon, C., & Teo, M. (2024). Artificial intelligence and sustainable development goals: Systematic literature review of the construction industry. *Sustainable Cities and Society, 108*, Article 105499. https://doi.org/10.1016/j.scs.2024.105499

Sage, J., & Pace, R. K. (2009). *Introduction to spatial econometrics*. CRC Press.

Sepehr, P. (2024). Mundane urban governance and AI oversight: The case of Vienna's intelligent pedestrian traffic lights. *Journal of Urban Technology*, Article 2302280. https://doi.org/10.1080/10630732.2024.2302280

Sharifi, A., Allam, Z., Bibri, S. E., & Khavarian-Garmsir, A. R. (2024). Smart cities and sustainable development goals (SDGs): A systematic literature review of co-benefits and trade-offs. *Cities, 146*, Article 104659. https://doi.org/10.1016/j.cities.2023.104659

Son, T. H., Weedon, Z., Yigitcanlar, T., Sanchez, T., Corchado, J. M., & Mehmood, R. (2023). Algorithmic urban planning for smart and sustainable development: Systematic review of the literature. *Sustainable Cities and Society, 94*, Article 104562. https://doi.org/10.1016/j.scs.2023.104562

Song, T., Dian, J., & Chen, H. W. (2023). Can smart city construction improve carbon productivity? A quasi-natural experiment based on China's smart city pilot. *Sustainable Cities and Society, 92*, Article 104478. https://doi.org/10.1016/j.scs.2023.104478

Song, M. L., Pan, H. T., Shen, Z. Y., & Tamayo-Verleene, K. (2024). Assessing the influence of artificial intelligence on the energy efficiency for sustainable ecological products value. *Energy Economics, 131*, Article 107392. https://doi.org/10.1016/j.eneco.2024.107392

Su, C. W., Yuan, X., Tao, R., & Shao, X. F. (2022). Time and frequency domain connectedness analysis of the energy transformation under climate policy. *Technological Forecasting and Social Change, 184*, Article 121978. https://doi.org/10.1016/j.techfore.2022.121978

Sun, H. P., Edziah, B. K., Sun, C. W., & Kporsu, A. K. (2019). Institutional quality, green innovation and energy efficiency. *Energy Policy, 135*, Article 111002. https://doi.org/10.1016/j.enpol.2019.111002

Tao, S. J., Wang, Y. Q., & Zhai, Y. N. (2023). Can the application of artificial intelligence in industry cut China's industrial carbon intensity? *Environmental Science and Pollution Research, 30*(33), 79571–79586. https://doi.org/10.1007/s11356-023-27964-5

Tian, M. X., Khan, K., Cifuentes-Faura, J., & Sukumaran, S. (2024). Technological innovation and energy efficiency in central Eastern European countries. *Utilities Policy, 88*, Article 101761. https://doi.org/10.1016/j.jup.2024.101761

Urom, C., Ndubuisi, G., Guesmi, K., & Benkraien, R. (2022). Quantile co-movement and dependence between energy-focused sectors and artificial intelligence. *Technological Forecasting and Social Change, 183*, Article 121842. https://doi.org/10.1016/j.techfore.2022.121842

Wang, S. J., Wang, J. Y., Fang, C. L., & Li, S. J. (2019). Estimating the impacts of urban form on CO_2 emission efficiency in the Pearl River Delta, China. *Cities, 85*, 117–129. https://doi.org/10.1016/j.cities.2018.08.009

Wang, E. Z., Lee, C. C., & Li, Y. Y. (2022). Assessing the impact of industrial robots on manufacturing energy intensity in 38 countries. *Energy Economics, 105*, Article 105748. https://doi.org/10.1016/j.eneco.2021.105748

Wang, Q., Li, Y. F., & Li, R. R. (2024). Ecological footprints, carbon emissions, and energy transitions: The impact of artificial intelligence (AI). *Humanities & Social Sciences Communications, 11*(1), Article 1043. https://doi.org/10.1057/s41599-024-03520-5

Wang, Y. Q., & Dong, W. (2020). How the rise of robots has affected China's labor market: Evidence from China's listed manufacturing firms. *Economics Research Journal, 55*(10), 159–175 (In Chinese).

Wei, X. H., Zhang, P. K., & Du, Y. H. (2020). How robots reshape the urban labor market: From a perspective of migrants' job tasks. *Economic Perspectives, 10*, 92–109. In Chinese.

Wei, G. E., He, B. J., Sun, P. J., Liu, Y. B., Li, R. Z., Ouyang, X., Luo, K., & Li, S. S. (2023). Evolutionary trends of urban expansion and its sustainable development: Evidence from 80 representative cities in the belt and road initiative region. *Cities, 138*, Article 104353. https://doi.org/10.1016/j.cities.2023.104353

Wong, I. A., Lian, Q. L., & Sun, D. N. (2023). Autonomous travel decision-making: An early glimpse into ChatGPT and generative AI. *Journal of Hospitality and Tourism Management, 56*, 253–263. https://doi.org/10.1016/j.jhtm.2023.06.022

Wu, J., Lin, K. X., & Sun, J. S. (2023). Improving urban energy efficiency: What role does the digital economy play? *Journal of Cleaner Production, 148*, Article 138104. https://doi.org/10.1016/j.jclepro.2023.138104

Yigitcanlar, T., Desouza, K. C., Butler, L., & Roozkhosh, F. (2020). Contributions and risks of artificial intelligence (AI) in building smarter cities: Insights from a systematic review of the literature. *Energies, 13*(6), Article 1473. https://doi.org/10.3390/en13061473

Yin, K. D., Cai, F. F., & Huang, C. (2023). How does artificial intelligence development affect green technology innovation in China? Evidence from dynamic panel data analysis. *Environmental Science and Pollution Research, 30*(10), 28066–28090. https://doi.org/10.1007/s11356-022-24088-0

You, J. S., Hu, J., & Jiang, B. (2025). Non-stationarity and spatial spillover effects in artificial intelligence development: Implications for sustainable urban transformation. *Sustainable Cities and Society, 131*, Article 106746. https://doi.org/10.1016/j.scs.2025.106746

Yu, J. H., Zhou, L. A., & Zhu, G. Z. (2016). Strategic interaction in political competition: Evidence from spatial effects across Chinese cities. *Regional Science and Urban Economics, 57*, 23–37. https://doi.org/10.1016/j.regsciurbeco.2015.12.003

Zeng, J., & Wang, T. (2025). The impact of China's artificial intelligence development on urban energy efficiency. *Scientific Reports, 15*(1), Article 24129. https://doi.org/10.1038/s41598-025-09319-x

Zeng, M., & Zhang, W. K. (2024). Green finance: The catalyst for artificial intelligence and energy efficiency in Chinese urban sustainable development. *Energy Economics, 139*, Article 107883. https://doi.org/10.1016/j.eneco.2024.107883

Zhang, R., & Fu, Y. H. (2022). Technological progress effects on energy efficiency from the perspective of technological innovation and technology introduction: An empirical study of Guangdong, China. *Energy Reports, 8*, 425–437. https://doi.org/10.1016/j.egyr.2021.11.282

Zhang, J. Y., & He, S. J. (2020). Smart technologies and urban life: A behavioral and social perspective. *Sustainable Cities and Society*, Article 102460. https://doi.org/10.1016/j.scs.2020.102460

Zhang, W. K., & Zeng, M. (2024). Is artificial intelligence a curse or a blessing for enterprise energy intensity? Evidence from China. *Energy Economics, 134*, Article 107561. https://doi.org/10.1016/j.eneco.2024.107561

Zhang, C. F., Dong, M. X., & Ota, K. R. (2020). Enabling computational intelligence for green internet of things: Data-driven adaptation in LPWA networking. *IEEE Computational Intelligence Magazine, 15*(1), 32–43. https://doi.org/10.1109/MCI.2019.2954642

Zhang, W. K., Fan, H. X., & Zhao, Q. W. (2023a). Seeing green: How does digital infrastructure affect carbon emission intensity? *Energy Economics, 127*, Article 107085. https://doi.org/10.1016/j.eneco.2023.107085

Zhang, W. K., Zeng, M., Zhang, Y. F., & Su, C. W. (2023b). Reducing carbon emissions: Can high-speed railway contribute? *Journal of Cleaner Production, 413*, Article 137524. https://doi.org/10.1016/j.jclepro.2023.137524

Zhang, G. Q., Chang, F. R., & Liu, J. (2024). Carbon emission prediction of 275 cities in China considering artificial intelligence effects and feature interaction: A heterogeneous deep learning modeling framework. *Sustainable Cities and Society, 114*, Article 105776. https://doi.org/10.1016/j.scs.2024.105776

Zhang, W. K., Fan, H. X., & Zeng, M. (2025). Unleashing the power of artificial intelligence: A game changer for urban energy efficiency in China. *Sustainable Cities and Society, 126*, Article 106372. https://doi.org/10.1016/j.scs.2025.106372

Zhao, H., Chen, S. W., & Zhang, W. K. (2023). Does digital inclusive finance affect urban carbon emission intensity: Evidence from 285 cities in China. *Cities, 142*, Article 104552. https://doi.org/10.1016/j.cities.2023.104552

Zhou, P., Ang, B. W., & Wang, H. (2012). Energy and CO_2 emission performance in electricity generation: A non-radial directional distance function approach. *European Journal of Operational Research, 221*, 625–635. https://doi.org/10.1016/j.ejor.2012.04.022

Zhou, W. N., Zhang, Y. Q., & Li, X. K. (2024). Artificial intelligence, green technological progress, energy conservation, and carbon emission reduction in China: An examination based on dynamic spatial Durbin modeling. *Journal of Cleaner Production, 446*, Article 141142. https://doi.org/10.1016/j.jclepro.2024.141142

Zou, W. Y., & Xiong, Y. J. (2023). Does artificial intelligence promote industrial upgrading? Evidence from China. *Economic Research-Ekonomska Istrazivanja, 36*(1), 1666–1687. https://doi.org/10.1080/1331677X.2022.2092168

Zulfiqar, M., Alshammari, N. F., & Rasheed, M. B. (2023). Reinforcement learning-enabled electric vehicle load forecasting for grid energy management. *Mathematics, 11*(7), Article 1680. https://doi.org/10.3390/mat21071680

CHAPTER 4

Urban Decarbonization Pathways of Digital Infrastructure

4.1 Introduction

Carbon emissions are a pressing concern that impacts the global community and represents an ongoing challenge that all nations must tackle earnestly. As the largest developing country, China faces imminent environmental issues, including carbon emissions. For instance, China's carbon emissions have grown rapidly from 2.173 billion tons in 1990 to 10.945 billion tons in 2020.[1] Overcoming the energy curse and achieving a low-carbon transformation are critical challenges not only for China but also for other countries striving for higher levels of development. China has actively pursued measures to lower carbon emissions (Gao et al., 2023; Lee et al., 2022a; Su et al., 2022b; Yahya & Lee, 2023; Zhang et al.,

[1] Data from the World Bank.

The core content of this chapter is derived from this article: Zhang, W. K., Fan, H. X., & Zhao, Q. W. (2023). Seeing green: How does digital infrastructure affect carbon emission intensity? *Energy Economics*, *127*, 107085. https://doi.org/10.1016/j.eneco.2023.107085

W. Zhang et al., *Green Transformation in the Digital Age*, https://doi.org/10.1007/978-981-95-9121-3_4

2023a, b). In 2020, China promised to achieve its carbon peak by 2030 and reach carbon neutrality by 2060. In this regard, China strongly advocates for the adoption of new environmentally friendly economic models, such as the digital economy, to optimize its industrial structure (Lee & Hussain, 2022; Hussain & Lee, 2022; Su et al., 2023; Yu et al., 2024; Zhang et al., 2023a, b), energy consumption (Zeng & Zhang, 2024; Zhang & Zeng, 2024; Zhang et al., 2025), and transportation (Avotra & Nawaz, 2023; Lee et al., 2022b; Zhang et al., 2023a, b; Zhao & Sun, 2022).

Digital infrastructure (DI) is a cutting-edge system driven by data innovation and built upon communication networks, with a focus on data computing power facilities (Tang & Yang, 2023; Yao et al., 2025; Zhang et al., 2023a, b). Chinese policymakers have increasingly recognized the significance of constructing DI in recent times. It has a positive influence on the promotion of industrial structure upgrades (Guo et al., 2022; Su et al., 2022a), drives innovation in business models (Ancillai et al., 2023), and enhances the efficiency of government market regulation (Rukanova et al., 2023). However, as DI continues to rapidly evolve and the ecological environment deteriorates, academia and industry have redirected their attention toward the impact of DI in environmental governance, particularly in reducing carbon emissions (Wang & Shao, 2024; Yi et al., 2024; Yu et al., 2025). In this context, the construction of DI assumes a critical and indispensable role. Therefore, it is imperative to delve deeper into the impact of DI on carbon emission intensity (CEI). Previous studies have extensively examined the environmental impacts of DI from various perspectives (e.g., Dong et al., 2022; Zhang et al., 2023a, b; Zheng et al., 2023). Some of these studies specifically have focused on particular DI, such as broadband China policy, information and communication technology (ICT), and digital transport infrastructure (Huong & Thanh, 2022). For example, Wu et al. (2021) employed Internet penetration as an indicator of DI to examine its impact on carbon emissions. Raheem et al. (2020) researched the influence of ICT on national-level carbon emissions. Nevertheless, little empirical literature has investigated in depth how DI influences CEI by constructing a comprehensive index, which is crucial for accurately assessing its environmental effects. More importantly, there is an ongoing debate regarding the effectiveness of enhancing DI in reducing carbon emissions. Some studies have concurred that DI is environmentally friendly and can contribute to carbon emission reduction (Madlener et al., 2022; Tao et al., 2025; Yi et al., 2022). However, other studies have argued that DI not only emits a significant amount of carbon dioxide during its construction process but also stimulates residents' consumption behavior, leading to increased energy consumption and exacerbating the current carbon emissions situation (Tang & Yang, 2023). At present, the task of carbon emission reduction is imminent, but the role of DI in CEI is still uncertain, and the existing literature has omitted a comprehensive assessment of DI.

This chapter aims to investigate the influence of DI on CEI by utilizing an unbalanced panel data of 275 Chinese cities from 2011 to 2019. The motivation for this chapter arises from several important aspects. Firstly, China, as one of the largest and fastest-growing emerging markets in the world, presents an intriguing research topic due to its cultural and historical background (Xie et al., 2019; Zhang et al., 2021). Secondly, different from developed economies in Europe and the United States, China faces significant challenges in reducing carbon emissions due to its large population and reliance on technology (Zhao et al., 2023). Thirdly, China boasts the world's largest and technologically advanced network infrastructure.[2] In 2022, the Chinese government introduced the "East Data and West Computing" policy, which seeks to establish a new computing power network system that integrates big data, data centers, and cloud computing to foster coordinated development between the eastern and western regions. Consequently, DI has emerged as a critical driver of economic development and a momentous element of national competitiveness in the modern era.

This chapter employs a city and year fixed effects model to identify the influence of DI on urban CEI. The findings reveal that DI can effectively reduce CEI in urban areas of China. Specifically, for every standard deviation increase of DI development level, there is a 1.9% reduction in CEI. These findings are robust and hold true even after conducting a series of additional tests. Additionally, a heterogeneity analysis is conducted to examine the variations in the carbon emission reduction of DI across geographical divisions (north-south divided by Qinling–Huaihe Line and east-west divided by Hu Huanyong Line), regional locations, and resource-based cities. The conclusions manifest that the influence of DI is more significant in northern cities, cities east of the Hu Huanyong Line, eastern and central cities, as well as resource-based cities. Furthermore, a mechanism analysis is performed to verify the channels through which

[2] https://www.miit.gov.cn/xwdt/gxdt/ldhd/art/2020/art_4dd5784bfbff487192b95ac92c12d437.html

DI affects CEI. This chapter identifies industrial upgrading, resource dependence, and green innovation as three important mechanisms. Finally, this chapter further explores the spatial spillover effects of DI on CEI by employing spatial econometric models. The findings offer preliminary evidence of a spatial spillover effect of DI on CEI.

4.2 Theoretical Analysis

4.2.1 The Effect of DI on CEI

DI refers to an innovative infrastructure system driven by data, centered around communication networks, and supported by data computing facilities (Ministry of Industry and Information Technology (MIIT, 2021). It primarily encompasses advanced information and communication technologies such as 5G, data centers, cloud computing, Internet of things, and blockchain. These technologies have given rise to various digital platforms that serve all aspects of people's work and daily lives. The impact of DI extends beyond work, as it also reshapes individuals' daily habits (Tang & Yang, 2023; Yang et al., 2022a). For instance, Fernández-Portillo et al. (2020) highlighted that promoting and developing information and communication technology (ICT) can significantly enhance production efficiency, thereby influencing employment and economic development. Tang and Yang (2023) proposed that the development of DI will revolutionize residents' behaviors, encouraging individuals to shape their consumption patterns and behaviors using digital devices. Additionally, Yang et al. (2022b) suggested that digitization will expand the scale of transportation and make individual transportation more convenient and efficient.

DI has the potential to dramatically change the economic development and energy consumption patterns of a country. However, its impact on carbon emissions remains a subject of controversy. Specifically, the influence of DI on urban CEI can be viewed from two perspectives: increasing or reducing CEI. On the one hand, the deployment of DI has the potential to intensify carbon emissions (Tang & Yang, 2023). Specifically, the construction and maintenance of DI require significant physical resources, including metals, plastics, and rare earths, resulting in substantial carbon emissions during the extraction and processing of these resources. Furthermore, DI inherently demands high energy consumption. The growth of the digital economy has also led to the widespread adoption of energy-intensive digital products and services, such as cloud computing, e-commerce, and online entertainment (Aldo et al., 2023). For instance, Aldo et al. (2023) highlighted how the utilization of DI has expanded the scale of online shopping, with the convenience of payment and transportation facilitating consumers' shopping habits, ultimately leading to increased energy consumption due to a sharp rise in shopping frequency. Colaco and de Abreu e Silva (2021) additionally argued that the promotion of DI not only encourages individuals to engage in excessive online shopping but also stimulates other consumers to consume beyond their needs through platform sharing, thereby exacerbating e-commerce usage. Tang and Yang (2023) contended that DI contributes to the escalation of urban carbon emissions by inducing higher per capita energy consumption.

On the other hand, DI plays a pivotal role in mitigating carbon emissions. This role is primarily demonstrated through DI's capacity to enhance energy efficiency, develop new clean resources, and incubate low-energy production and lifestyles. Firstly, DI enhances energy efficiency by leveraging digital technology for intelligent management and optimization of energy systems. An exemplary application is the utilization of smart grid systems, which employ cloud computing technology to accurately predict and schedule power demand, optimize the power supply chain, minimize energy waste, and thus reduce carbon emissions (Bakken, 2019). Secondly, DI facilitates the development and utilization of renewable energy sources, thereby reducing reliance on conventional energy and decreasing carbon emissions. For instance, smart power systems based on big data and artificial intelligence technology enable precise prediction and scheduling of renewable energy, enhance its utilization efficiency, lessen the demand for traditional energy sources, and subsequently lower carbon emissions (Li et al., 2019). Yang et al. (2022b) found that DI overcomes regional and temporal constraints, accelerates the production and flow of factors, saves energy waste caused by production and lifestyle obstacles, improves energy utilization efficiency, and reduces carbon emissions. Finally, DI drives the advancement of electric transportation, smart cities, remote working, and distance learning, ultimately curbing energy consumption and carbon emissions. For example, establishing a network of charging stations and implementing intelligent transportation systems promote the convenience and reliability of electric vehicles, thereby decreasing the employ of conventional fuel vehicles and lowering carbon emissions in urban transportation. Similarly, the development of DI facilitates telecommuting and distance learning, reducing the need for daily commuting and resulting in a decrease in carbon emissions from urban transportation while enhancing resource utilization efficiency.

In general, the carbon emissions associated with the construction of DI are temporary. As the DI technology and its operation become more mature, the carbon emissions from DI itself become limited (Yang et al., 2022a). However, it is important to note that DI has a persistent influence on reducing CEI. Consequently, the overall impact of DI on CEI is more likely to result in a reduction rather than an increase.

This chapter proposes the following hypothesis:

Hypothesis 4.1: DI can reduce urban CEI.

4.2.2 Heterogeneous Impact of DI on CEI

China's expansive geographical expanse and intricate social attributes give rise to notable disparities among cities in terms of physical geography, economic development, cultural dynamics, and scientific and technological advancements. Consequently, investigating the varying impacts of DI on urban CEI has emerged as a central focus in understanding this heterogeneity. Specifically, this chapter discusses the differential effects within cities located in the north-south divided by Qinling–Huaihe Line, the east-west divided by Hu Huanyong Line, regional location, and resource-based cities.

The Qinling–Huaihe Line is widely acknowledged as the boundary separating the northern and southern regions of China. These two regions exhibit distinct disparities in terms of terrain, industry composition, resource allocation, economic development, technological foundation, as well as political and cultural traditions (Zhang et al., 2023a, b). These disparities will inevitably result in varying degrees of impact from DI on CEI. The northern region is predominantly characterized by heavy industries such as steel, metallurgy, machinery, petroleum, and chemistry, consequently leading to significantly higher carbon emissions compared to the southern region, which is primarily focused on light industry. Moreover, the southern region possesses abundant water resources and economic development exhibits lower dependence on coal and oil in comparison to the northern region. Energy consumption, a key determinant of carbon emissions (Wang & Shao, 2023), is higher in the northern region than in the southern region (Dong et al., 2018b). On the whole, the space for energy consumption reduction in the northern region is larger than that in the southern region, while the energy consumption in the southern region is less and the room for decline is less. DI has the potential to enhance the efficiency of these traditional energy sources (Haldar & Sethi, 2022). As a result, the impact of DI on CEI is likely to be more pronounced in northern cities than in their southern counterparts.

This chapter proposes the following hypothesis:

Hypothesis 4.2: The impact of DI on CEI is more significant in northern cities bounded by the Qinling–Huaihe Line than in southern cities.

The Hu Huanyong Line serves as a significant geographical demarcation, with notable variations in population densities on its eastern and western sides (Wang & Deng, 2016). For instance, more than 90% of the country's residents are concentrated on just 30% of the land to the east of this line, exhibiting significantly higher population density compared to the western cities. The population factor plays a vital role in shaping the extent of DI. During the initial stages of DI construction, several environmental issues arise, leading to an increase in regional CEI. However, as the DI matures, the urban CEI decreases (Yang et al., 2022a). Furthermore, the population factor significantly influences the development level of DI. The eastern part of the Hu Huanyong Line possesses a greater advantage in terms of human resources, resulting in faster development of DI and a more pronounced effect on CEI.

This chapter proposes the following hypothesis:

Hypothesis 4.3: The impact of DI on CEI is more pronounced in eastern cities bounded by the Hu Huanyong Line than in western cities.

China's expansive territory encompasses significant variations in resource endowment, economic scale, scientific and technological advancement, and openness in eastern, central, and western cities. Consequently, the development level of DI varies across these cities. The eastern and central cities benefit from advantageous geographical locations, higher levels of economic development, abundant human capital, and efficient transportation systems, which distinguishes them from their western counterparts (Zhang et al., 2023a, b). Furthermore, the predominantly flat terrain in the eastern and central cities facilitates the implementation of DI (Hong et al., 2023). In comparison, the western region exhibits a more dispersed industrial distribution and lower energy consumption (Dong et al., 2018a). The central and eastern cities share close interconnections, and central cities are heavily influenced by the development of the eastern region, resulting in a relatively rapid pace of DI. Conversely, the western cities face challenges due to their undulating terrain, sparse population, and underdeveloped economic conditions, which hinder the construction of DI. Moreover, industries in the western cities are scattered and lack scalability, leading to a relatively smaller carbon emissions base. The limited proximity and varying terrain and climate conditions across regions restrict the potential of DI to contribute to energy conservation and carbon reduction.

This chapter proposes the following hypothesis:

Hypothesis 4.4: The impact of DI on CEI is more pronounced in central and eastern cities than in western cities.

Resource-based cities primarily rely on the extraction and processing of resources (Che & Wang, 2022). These cities have established energy industries and experienced significant economic growth, but they are also accompanied by poor environmental conditions. Cities are categorized into resource-based and non–resource-based cities based on their relationship with traditional resources. These two types of cities are significantly different in terms of development policies, economic levels, industrial structures, and other aspects. Resource-based cities heavily rely on the extraction of resources, resulting in the phenomenon known as the "resource curse" (Auty, 2002). This overreliance on natural resources has led to a dilemma of environmental conditions and economic development. However, the construction of DI has presented new opportunities for these traditional resource-based cities. It has allowed them to quickly seize opportunities and explore alternative models of economic development, while also offering prospects for environmental improvement (Zhang et al., 2022b; Wang et al., 2023).

This chapter proposes the following hypothesis:

Hypothesis 4.5: The impact of DI on CEI is more pronounced in resource-based cities than in non–resource-based ones.

4.2.3 The Mechanisms of the Impact of DI on CEI

The industrial structure plays a pivotal role in driving economic growth, but it also contributes significantly to carbon emissions. The current state of the industrial sector profoundly influences the level of carbon emissions (Lin & Wang, 2023; Cheng et al., 2018). Upgrading the industrial structure involves adopting more scientific production methods and optimizing the production processes, thereby reducing unnecessary energy consumption and effectively mitigating CEI (Wang et al., 2023). On the one hand, DI plays a vital role in facilitating the upgrading of the industrial structure. By providing convenience for enterprises to integrate advanced production technologies, it facilitates the optimization of the existing production structure. Additionally, it streamlines information exchange, allowing for increased investment in more advanced industries (Gong et al., 2023). Furthermore, DI empowers consumers to elevate their consumption patterns, thereby pressuring the transformation and upgrading of existing industries (Ren et al., 2023). On the other hand, DI also exerts a pulling force on industrial structure upgrading by enhancing production efficiency. This enables advanced industries to generate higher profits, thereby incentivizing enterprises to actively explore avenues for upgrading (Jiang et al., 2023).

This chapter proposes the following hypothesis:

Hypothesis 4.6: DI reduces CEI through the upgrading of the industrial structure.

Previous studies have observed that cities characterized by a higher degree of reliance on natural resources are more prone to engaging in more frequent resource development activities, which in turn leads to an increase in carbon emissions (Wang et al., 2019). Moreover, it has been suggested that cities abundant in natural resources may face distorted resource development and utilization, leading to elevated carbon emissions. DI plays a crucial role in transforming traditional manual operations into precise assembly line processes, thereby minimizing energy waste significantly (Hong et al., 2023). This transformation enables enterprises to adapt more effectively to different elements, enhancing the efficiency of energy utilization and reducing reliance on traditional natural resources. Additionally, DI can optimize the energy consumption structure, consequently reducing cities' dependence on energy (Ren et al., 2021).

This chapter proposes the following hypothesis:

Hypothesis 4.7: DI reduces CEI by decreasing cities' reliance on natural resources.

The consensus among academic researchers is that innovation, particularly in green technology, plays a vital role in expediting urban energy conservation and carbon reduction (Zhang & Liu, 2022; Zhao et al., 2023). The utilization of DI facilitates communication and collaboration among innovation entities, thereby providing convenient access to innovation resources (Hong et al., 2023). Essentially, DI offers advantages such as low transmission costs and rapid information dissemination, effectively promoting the diffusion of existing technologies and the emergence of new ones (Xu et al., 2019). Moreover, DI strengthens the government's monitoring capabilities and compels enterprises to engage in green technology innovation through enhanced supervision (Li et al., 2021).

This chapter proposes the following hypothesis:

Hypothesis 4.8: DI reduces CEI by enhancing green innovation of cities.

4.3 Research Design

4.3.1 Data

This chapter employs an unbalanced panel dataset comprising 275 cities from 2011 to 2019 to investigate the impact of DI on urban CEI in China. The CEI data is sourced from Carbon Emission Accounts and Datasets (CEADs) for emerging economies released by Tsinghua University, while the DI data is collected from the Chinese Research Data Services (CNRDS) platform (www.cnrds.com). Additional city-level data is sourced from the China Cities Statistical Yearbook, and information on innovative and resource-based cities is retrieved from China's official government website. To account for the unique situation of municipalities directly under the central government, the samples of Tianjin, Beijing, Shanghai, and Chongqing are excluded during the processing stage. The final dataset consists of 2474 city–year observations.

4.3.2 Variable

The explained variable The independent variable in this chapter is CEI. To calculate CEI of each prefecture-level city in China, this chapter utilizes the calculation methods developed by Wang et al. (2019, 2022), and Zhang et al. (2023a, b), which involve taking the natural logarithm of carbon emissions per unit of GDP. The specific calculation formula is as follows:

$$CEI_{it} = \ln \frac{CE_{it}}{GDP_{it}} \tag{4.1}$$

where i represents each prefecture-level city, t represents year, CE is the total carbon emissions, GDP represents the regional gross domestic product, CEI indicates carbon emission intensity.

The explanatory variable The independent variable of this chapter is DI. In line with the previous studies of Zhao and Sun (2022), Fan et al. (2022), and Zhang et al. (2023a, b), this chapter extracts four dimensions to construct the DI index. These dimensions encompass broadband Internet base, mobile Internet base, telecommunications base, information and software scale. Specifically, the broadband Internet base is measured by the number of Internet broadband access users, the mobile Internet base is calculated by the number of mobile phone users, the telecom base is measured by telecom business income, and the information and software scale is measured by information transmission computer service and software industry personnel. The construction of the DI index in this chapter is based on the principal component analysis (PCA) method. Furthermore, to enhance the reliability of the results, this chapter performs robustness tests by employing DI calculated through the factor analysis method (DI_FA) and digital financial inclusion (DFI) index proposed by Peking University.

The control variables Referring to Tang and Yang (2023) as well as Zhang et al. (2023a, b), this chapter further controls for other potential factors that may affect CEI. Specifically, these control variables include economic development, energy consumption, population density, industrial structure, technical level, and educational standard. Economic development is calculated as the growth rate of GDP. Energy consumption is determined by dividing coal consumption by total energy consumption. Education standard is measured by taking the natural logarithm of per capita education expenditure. Population density is calculated by taking the logarithm of the population per unit area of the city. Industrial structure is estimated by the ratio of the secondary industry output value to GDP. Technical level is measured by taking the natural logarithm of per capita financial expenditure on science and technology.

4.3.3 Model

This chapter employs a city–year fixed effects model to capture the influence of DI on CEI in China, and constructs as follows:

$$CEI_{it} = \alpha_0 + \beta_1 DI_{it} + \beta_2 \text{Controls}_{it} + \mu_i + \gamma_t + \varepsilon_{it} \tag{4.2}$$

where CEI represents urban CEI; DI represents the comprehensive development level of DI; β_1 is the coefficient of interest DI; i and t represent city and year, respectively; *Controls* represents control variables, and β_2 denotes their corresponding coefficients; μ_i and γ_t separately represent city and year fixed effects; ε_{it} represents a random error vector. Table 4.1 presents the descriptive statistics of the main variables.

Table 4.1 Descriptive statistics

Variables	*Observations*	*Mean*	*S.D.*	*Min*	*Max*
CEI	2474	2.698	0.566	0.163	5.445
DI (PCA)	2474	1.561	0.727	0.618	7.599
DI_FA (Factor analysis)	2474	1.580	0.729	0.338	9.697
DFI	2474	167.304	67.481	17.020	326.767
Energy consumption	2474	0.730	0.313	0.248	2.461
Industrial structure	2474	47.187	10.532	11.700	89.340
Population density	2474	10.353	0.894	6.225	12.528
Economic development	2474	8.693	3.767	−19.380	37.540
Technological level	2474	4.462	1.149	1.750	9.409
Educational standard	2474	7.210	0.453	4.159	9.473

Table 4.2 Benchmark regression

Variables	*(1)*	*(2)*	*(3)*
DI	−0.028***	−0.019***	−0.019**
	(0.007)	(0.005)	(0.007)
Constant	2.741***	5.976***	5.976***
	(0.011)	(0.919)	(1.721)
Controls	No	Yes	Yes
City FE	Yes	Yes	Yes
Year FE	Yes	Yes	Yes
Observations	2474	2474	2474
R-squared	0.943	0.963	0.963

Robust or clustered standard errors in parentheses; *** $p < 0.01$, ** $p < 0.05$, * $p < 0.10$

4.4 Findings

4.4.1 Baseline Results

Table 4.2 presents the benchmark results. Columns (1) and (2) exhibit the estimations before and after controlling for other variables, while considering robust standard errors. Column (3) displays the estimates with all control variables clustered at the city level.

The results reveal that, regardless of the inclusion of control variables, the estimates of DI are remarkably negative at a 5% level. This suggests that DI has a substantial and adverse impact on CEI. This finding confirms Hypothesis 4.1 and demonstrates the effectiveness of DI in reducing urban CEI in China. Moreover, the coefficients for DI in Columns (2) and (3) are −0.019, indicating a 1.9% decrease in CEI for every standard deviation increase in the level of DI construction.

These findings can be attributed to the influence of DI on the behavior of governments, businesses, and residents, resulting in a reduction in urban CEI. First, improved DI construction enables governments to efficiently and accurately obtain carbon emission-related data and formulate timely response measures (Alam & Murad, 2020; Zhang et al., 2022c). Second, DI benefits enterprises by optimizing resource allocation and upgrading production processes, leading to carbon emissions reduction (Martínez et al., 2022). Finally, digital networks facilitate low-carbon consumption by reshaping consumers' awareness of low-carbon practices and improving their consumption patterns (Nudurupati et al., 2022).

The aforementioned findings align with the conclusions reached by Lu (2018), Raheem et al. (2020), and Li et al. (2025), who investigated the influence of ICT on carbon emissions at the national level. Lu's study revealed a remarkable negative correlation between ICT and carbon dioxide emissions, whereas Raheem's findings indicated that this negative effect only becomes evident in the long term. Moreover, Hu et al. (2023) explored the influence of DI construction on low-carbon development and proposed that DI can directly facilitate urban low-carbon development. This reference differs from our study as it primarily focuses on low-carbon development proxied by carbon total factor productivity. However, our study contradicts the conclusions of Tang and Yang (2023). Tang and Yang (2023) employed the data of prefecture-level cities to examine the influence of DI on carbon emissions, but they concluded that DI positively increases urban carbon emissions. It is noteworthy that Tang and Yang (2023) utilized the number of keywords related to DI in government documents as the evaluation index for DI. This approach may result in a policy lag, making it challenging to accurately evaluate the impact of DI on carbon emissions.

4.4.2 Robustness Test

Alternative measures for CEI This chapter conducts robustness tests using two alternative measures for CEI. The first alternative measure replaces the original strategy with data from the Multi-resolution Emission Inventory for China (MEIC) database, which is published by Tsinghua University. The MEIC database provides high-resolution air pollutant and carbon emission datasets. Since the database only includes carbon emission data of provincial-level cities, this chapter calculates total carbon emissions for prefecture-level cities by considering the proportion of each city's population within its respective province. This approach helps reduce bias arising from single data statistics. The estimation is then conducted using Eq. (4.2), and the results are reported in Column (1) of Table 4.3. The findings indicate that the estimate of DI is still negative at a 1% significance level, further supporting the notion that DI significantly contributes to the reduction of CEI.

Another alternative approach replaces CEI with per capita carbon emissions. Considering the significant differences in population size among cities, this approach helps mitigate potential bias in the analysis results.

Table 4.3 Robustness test results

Variables	*Alternative measures for CEI*		*Alternative measures for DI*		*Eliminate policy interference*	
	(1)	*(2)*	*(3)*	*(4)*	*(5)*	*(6)*
DI	−0.016*** (0.003)	−0.002*** (0.001)			−0.021*** (0.005)	−0.019*** (0.005)
DI_FA			−0.009* (0.005)			
DFI				−0.003*** (0.001)		
High-speed rail						−0.006 (0.009)
Constant	−0.003 (0.337)	1.362*** (0.279)	6.135*** (0.924)	5.850*** (0.888)	5.424*** (1.070)	5.988*** (0.920)
Controls	Yes	Yes	Yes	Yes	Yes	Yes
City FE	Yes	Yes	Yes	Yes	Yes	Yes
Year FE	Yes	Yes	Yes	Yes	Yes	Yes
Observations	2474	2474	2474	2474	2186	2474
R-squared	0.940	0.988	0.963	0.964	0.959	0.963

Robust standard errors in parentheses; *** $p < 0.01$, ** $p < 0.05$, * $p < 0.10$

Hence, per capita carbon emissions are chosen as a substitute for carbon emissions per unit of GDP. The results, presented in Column (2) of Table 4.3, demonstrate that the estimate of DI remains notably negative at a 1% level. These results reconfirm the reliability of the above results in benchmark regression.

Alternative measures for DI This section incorporates two alternative measures for DI to perform robustness tests. Firstly, in the benchmark regression, the DI index is derived by applying the PCA method. Recognizing the potential discrepancies that may arise from using different calculation techniques, this analysis employs the DI index generated via factor analysis (DI_FA) as an additional robustness test. Secondly, this chapter calculates the DI using DFI proposed by Peking University. Columns (3) and (4) of Table 4.3 present the results respectively. Importantly, both coefficients demonstrate significant negativity at the traditional statistics levels, consistent with the benchmark results, thus confirming the accuracy of the aforementioned findings.

Eliminate policy interference This chapter initially considers the pilot projects aimed at reducing carbon emissions, which may influence the values of carbon emissions and cause deviations from the regression results. Specifically, in 2011, several local pilot projects for carbon emissions trading were implemented in Shanghai, Chongqing, Tianjin, Beijing, Guangdong, and Hubei. Consequently, these cities have been excluded from the regression analysis in this chapter, and the results are reported in Column (5) of Table 4.3. The estimate of DI remains significantly negative at a 1% level, providing further confirmation that the influence of DI on CEI reduction is not affected by other relevant policies.

Moreover, previous studies have indicated that the introduction of high-speed rail can decrease urban carbon emissions (Zhang et al., 2023a, b). Therefore, this chapter includes the variable of whether a city has a high-speed rail system as a controlling factor. The results in Column (6) of Table 4.3 indicate that the estimate of DI is still observably negative at a 1% level. This finding further confirms the robustness of the aforementioned conclusions.

Endogenous analysis Despite controlling for several crucial variables that influence urban CEI and using a year and city fixed-effects model to mitigate estimation bias, this chapter may not fully resolve the issue of endogeneity. Additionally, the inverse causality problem is considered insignificant since urban carbon emissions typically do not have a significant impact on digital technology facilities. To address potential endogeneity resulting from others, this chapter employs the instrumental variable method to perform the endogeneity analysis.

The first strategy involves utilizing the number of fixed telephones per 100 individuals in 1984 as an instrumental variable for DI (*DI_IV*) to conduct an endogeneity analysis. The selection of this instrumental variable is based on several reasons. Firstly, the development of traditional communication technology, as represented by fixed telephones, has historically influenced the construction of DI in terms of technical level and usage patterns. Secondly, the influence of traditional telecommunication tools such as fixed-line phones on carbon emissions gradually decreases as their usage frequency declines, satisfying the requirement of exclusivity (Huang et al., 2019). Since the above instrumental variable is cross-sectional data, this chapter employs its interaction term with the number of Internet broadband access users as an instrumental variable and applies the 2SLS method for regression analysis. The results, presented in Column

Table 4.4 Instrumental variable results

Variables	*Number of fixed telephones in 1984*	*One-period-lagged DI*
	(1)	*(2)*
DI_IV	−0.040***	
	(0.015)	
DI_lag		−0.044***
		(0.012)
Controls	Yes	Yes
City FE	Yes	Yes
Year FE	Yes	Yes
Observations	1952	2198
R-squared	0.399	0.332
F-value for the first stage	330.100	587.410

Robust standard errors in parentheses; *** $p < 0.01$, ** $p < 0.05$, * $p < 0.10$

(1) of Table 4.4, indicate that the F-statistics value for the first stage exceeds 10, suggesting that there are no weak instrumental variable problems. The coefficient of *DI_IV* is significantly negative at a 1% level, consistent with the aforementioned findings. These results provide additional evidence supporting the effectiveness of DI in reducing CEI in China.

The second strategy proposed by Hu et al. (2023) involves using one-period-lagged DI as an instrumental variable. By using one-period-lagged DI, a direct association with the current period is ensured, while avoiding any direct link to the CEI of the current period. This approach meets the requirements of exogeneity. The results are reported in Column (2) of Table 4.4. The corresponding F-statistics value of the first stage exceeds 10, suggesting that the instrumental variables, represented by the one-period-lagged DI, is not weakened. These results are in line with the aforementioned results and provide additional evidence of the effectiveness of DI in reducing CEI in China.

4.4.3 *Heterogeneity Analysis*

North-south divide Building upon the research conducted by Zhang et al. (2023a, b), this chapter categorizes China's cities into two regions, namely north and south, with the Qinling–Huaihe Line serving as the demarcation. The objective of this chapter is to investigate the heterogeneous impact of DI on CEI within these distinct regions. It is crucial to consider the varying natural conditions between the two regions, as this is a critical factor when assessing the relationship between DI and CEI. The results, presented in Columns (1) and (2) of Table 4.5, suggest that the coefficient of DI in northern cities is negative at a 1% level of significance. This indicates that DI has a remarkable capability to reduce CEI in northern cities. In contrast, the coefficient in southern cities is negative but not statistically significant. This implies that DI may not exert a noticeable influence on CEI in southern cities.

The reason for the above disparities between southern and northern cities can be primarily attributed to their distinct natural geographical environments. On the one hand, these differences stem from the varying terrains in the two types of cities. The northern cities are predominantly flat, which facilitates the dissemination of DI. Conversely, the southern cities comprise complex terrains, including mountains and hills, which inevitably pose numerous challenges to the construction of DI. Consequently, the obstacles encountered in the construction of DI in northern cities are relatively minimal, resulting in faster growth in terms of both speed and scale. On the other hand, the dissimilar natural conditions between the north and south also give rise to disparities in their industrial structures. Historically, heavy industries with high energy consumption have dominated northern cities, consequently making the northern cities the most severely affected area in terms of carbon emissions. The development of DI can optimize the traditional heavy industrial structure in the north and introduce an environmentally friendly economic development model, thereby mitigating the pressure of carbon emissions.

East and west divide This section utilizes the Hu Huanyong Line, a demographic boundary proposed by professors from Ohio State University in the United States, to partition cities into east and west for the purpose of investigating the influence of DI on CEI. This line serves as a significant geographical demarcation widely recognized and applied in academic research. It categorizes cities into east and west regions, with the east being densely populated and the west being sparsely populated. The estimated values for cities on both sides of Hu Huanyong Line are displayed in Columns (3)–(4) of Table 4.5.

The results suggest that the coefficient of DI in eastern cities along the Hu Huanyong Line is noticeably negative at a 1% level, with a value of −0.024. However, the coefficient for cities in the west is insignificant and negative at the traditional statistical levels. These findings indicate that

Table 4.5 Regional heterogeneity

Variables	*North-south divided by Qinling–Huaihe line*		*East-west divided by Hu Huanyong line*		*Regional location*		
	Southern	*Northern*	*Hu-eastern*	*Hu-western*	*Eastern*	*Central*	*Western*
	(1)	*(2)*	*(3)*	*(4)*	*(5)*	*(6)*	*(7)*
DI	−0.006	−0.024***	−0.024***	0.002	−0.024*	−0.032**	−0.005
	(0.007)	(0.007)	(0.005)	(0.016)	(0.012)	(0.013)	(0.006)
Constant	2.539**	7.531***	5.301***	6.828***	7.331***	4.166***	6.508***
	(1.238)	(1.319)	(0.918)	(1.861)	(1.295)	(1.514)	(2.016)
Controls	Yes	Yes	Yes	Yes	Yes	Yes	Yes
City FE	Yes	Yes	Yes	Yes	Yes	Yes	Yes
Year FE	Yes	Yes	Yes	Yes	Yes	Yes	Yes
Observations	1295	1179	2267	207	881	945	648
R-squared	0.965	0.943	0.961	0.963	0.961	0.956	0.976

Robust standard errors in parentheses; *** $p < 0.01$, ** $p < 0.05$, * $p < 0.10$

DI plays an observable role in reducing CEI in eastern cities along the Hu Huanyong Line, while its effect on the CEI in cities to the west of the line is not evident.

One possible explanation for this phenomenon is the disparity in demand and supply of DI, which is influenced by significant demographic and geographic differences on both sides of the line. The cities located in the eastern part of the line are characterized by narrow spaces and high population density, accounting for over 90% of the total population. Conversely, the west of the line is characterized by vast expanses of land with sparse population distribution. The construction of DI is closely tied to the distribution of population. On the one hand, the construction of DI primarily aims to satisfy the needs and facilitate the development of the population. The densely populated cities located east of the line have a substantial demand for DI, which motivates them to invest in its development. On the other hand, the construction of DI necessitates a skilled workforce, including technical experts and grassroots workers. In this regard, the cities located east of the line have a distinct advantage. Consequently, the development level of DI is higher in the east compared to in the west, and the impact of DI on reducing CEI is more pronounced in densely populated cities located east of the line.

Regional location This section categorizes the sample into central, eastern, as well as western cities based on the criteria set by the National Bureau of Statistics of China, in order to investigate the impact of DI on CEI. The heterogeneity results of regional location are presented in Columns (5)–(7) of Table 4.5.

The estimated coefficients of DI in eastern and central cities exhibit significant negative values at conventional statistical significance levels, specifically −0.024 and −0.032. However, the coefficient for western cities is negative but fails to reach statistical significance at the conventional levels. These findings suggest that DI in eastern and central cities has a significant capability to reduce CEI, whereas the impact of DI in western cities on CEI is not statistically significant. This finding aligns with the research conclusion of Zhang et al. (2022a).

The heterogeneous results mentioned above can be attributed to several factors. Firstly, there exist disparities in terms of economic base, industrial structure, and technological level between cities in the central and eastern regions and those in the western regions. Secondly, the central and eastern regions possess a strong economic foundation and demonstrate significant capabilities in scientific and technological innovation capabilities. Additionally, the speed and level of DI construction in these regions are expected to surpass those of the western region. Lastly, the implementation of DI in central and eastern regions can facilitate the upgrading of their industrial structure by eliminating outdated production capacity and potentially relocating industries with high energy consumption and pollution to the western region. Consequently, this shift could mitigate CEI in central and eastern regions while simultaneously increasing the pressure on carbon emissions in the western region.

Resource-based city Resource-based cities are defined by their dominant industries in mining and processing natural resources, including minerals and forests. The National Sustainable Development Plan for Resource-Based Cities (2013–2020) classifies cities into two categories: resource-based cities and non–resource-based cities, based on their industrial structure and economic growth. To investigate the impact of DI on CEI, this chapter further subdivides the sample into resource-based and non–resource-based categories. The results are presented in Columns (1)–(2) of Table 4.6.

The findings indicate that the estimated coefficient of DI in resource-based cities is prominent negative at a 1% level, with a coefficient of −0.029. However, it is not significant at conventional levels for non–resource-based cities. This suggests that the influence of DI on CEI is

Table 4.6 The results of resource-based city

Variables	*Resource--based*	*Non–resource--based*	*Growing*	*Mature*	*Declining*	*Regenerative*
	(1)	*(2)*	*(3)*	*(4)*	*(5)*	*(6)*
DI	−0.029***	−0.011	−0.003	−0.029**	−0.032	−0.055*
	(0.008)	(0.007)	(0.012)	(0.013)	(0.034)	(0.031)
Constant	6.873***	5.574***	10.735*	5.722**	6.783**	14.173***
	(1.844)	(1.083)	(5.513)	(2.834)	(3.113)	(5.201)
Controls	Yes	Yes	Yes	Yes	Yes	Yes
City FE	Yes	Yes	Yes	Yes	Yes	Yes
Year FE	Yes	Yes	Yes	Yes	Yes	Yes
Observations	990	1484	117	549	207	135
R-squared	0.942	0.972	0.960	0.967	0.954	0.818

Robust standard errors in parentheses; *** $p < 0.01$, ** $p < 0.05$, * $p < 0.10$

more pronounced in resource-based cities. One possible explanation for these findings is that resource-based cities are more dependent on energy-intensive and highly polluting industries associated with traditional resources, resulting in higher carbon emissions and more severe environmental issues. DI can play a crucial role in helping resource-based cities upgrade their existing industrial models, eliminate outdated production capacities, and reduce carbon emissions. Additionally, DI can facilitate the introduction of cleaner industries to resource-based cities and promote the transformation of their development models.

Besides, this chapter investigates the heterogeneity of resource-based cities at different stages of growth by categorizing them into growing, mature, declining, and regenerative. Columns (3)–(6) of Table 4.6 display the specific findings. The estimates of DI in mature and regenerative cities are found to be significantly negative at 10% or 5% significance levels, with values of −0.029 and −0.055, respectively. However, in growing and declining cities, these coefficients are negative but not statistically significant. These results suggest that DI has a more observable impact on reducing CEI in mature resource cities and renewable resource cities.

The possible reason for these findings is that mature cities often possess a relatively strong economic foundation, enabling them to invest more in the development of DI. As a result, they tend to have a higher level of DI. Regenerative cities have managed to reduce their dependence on resources and have experienced positive economic development. These cities represent the forefront of economic transformation among cities and generally exhibit a high level of DI, thereby leading to a more noticeable reduction in CEI. In contrast, growing cities are in the early stages of resource development, characterized by relatively weak economic and technological capabilities and an overall lag in DI. Similarly, declining cities face challenges such as resource depletion, economic stagnation, and prominent livelihood issues, making it difficult for them to allocate sufficient resources to promote the development of DI.

4.4.4 Underlying Mechanisms

Industrial upgrading Existing studies have verified that the upgrading of industrial structure is effective in reducing carbon emissions (Yang et al., 2022b; You & Zhang, 2022). As suggested by Chen et al. (2020), to test the feasibility of this underlying mechanism, this chapter makes a point of investigating the role of DI in upgrading the industrial structure.

Table 4.7 Underlying mechanism results

Independent variables	*Industrial upgrading*	*Resources dependence*	*Green innovation*
	(1)	*(2)*	*(3)*
DI	0.490***	−0.002**	0.033**
	(0.170)	(0.001)	(0.014)
Constant	284.780***	−0.998***	−2.334
	(18.147)	(0.230)	(2.108)
Controls	Yes	Yes	Yes
City FE	Yes	Yes	Yes
Year FE	Yes	Yes	Yes
Observations	2474	2002	2468
R-squared	0.971	0.962	0.962

Robust standard errors in parentheses; *** $p < 0.01$, ** $p < 0.05$, * $p < 0.10$

Specifically, the industrial upgrading index is calculated by assigning weights of 1, 2, and 3 to the primary, secondary, and tertiary industries, respectively. A higher value indicates a more significant industrial upgrading. The estimation of industrial upgrading is presented in Column (1) of Table 4.7. The results evidence a significant coefficient of 0.490 for DI in industrial upgrading at a 1% level. This suggests that DI plays a vital role in accelerating the industrial structure upgrading. Consequently, it is evident that DI contributes to CEI reduction by facilitating industrial structure upgrading.

The potential reason is that DI can effectively enhance production efficiency across multiple sectors by coordinating production and consumption activities. It facilitates the optimization of industrial factors allocation, strengthens market competition mechanisms, and supports the incubation of emerging ecological industries. It also eliminates outdated production capacity and promotes industrial structure upgrading. From the production perspective, DI optimizes the industrial chain by breaking down information barriers between production departments, producers, and consumers. This closer connection improves productivity and reduces energy waste. On the consumer side, DI not only provides convenience and speed in shopping but also encourages consumers to pursue personalized and environmentally friendly products. This trend stimulates the original industry to improve production methods to cater to consumer preferences and promotes the emergence of new low-energy consumption industries.

Resources dependence Several scholars have suggested that the phenomenon known as the "resource curse," which arises from cities excessively relying on resources, not only hinders the rapid development of urban economies but also obstructs low-carbon development (Abraham, 2021; Che & Wang, 2022; Li et al., 2022). Mitigating the resource curse and decreasing reliance on traditional resources represent effective strategies for reducing carbon emissions. Consequently, this chapter initially quantifies resource dependence by assessing the proportion of coal industry employees among those employed in the secondary industry. Subsequently, it evaluates the impact of DI on resource dependence.

The estimation of resource dependence is presented in Column (2) of Table 4.7. It is evident that the estimate of DI on resource dependence is −0.002 at a 5% significant level. This finding contends that DI has the potential to reverse the excessive reliance of cities on resources, thereby breaking the cycle of resource dependence and improving the urban environmental situation. On the one hand, DI can promote the adoption of new technologies and business models, reducing cities' dependence on traditional natural resources and facilitating energy conservation and emissions reduction. On the other hand, DI facilitates the communication of information and technology between cities and the external world, enabling the introduction of new clean resources for urban development and mitigating urban carbon emissions.

Green innovation Green innovation plays a crucial role in the endeavor to reduce carbon emissions (Zhao et al., 2023). This section further explores the potential of DI in lowering the CEI of urban areas through the promotion of urban green innovation. This chapter measures green innovation by applying the natural logarithm of green patent applications within each prefecture-level city, and the estimation is reported in Column (3) of Table 4.7.

The findings reveal that DI has a significant positive impact on green innovation, with an estimated coefficient of 0.033 at a 5% significance level. This provides strong evidence that DI plays a dominant role in advancing urban green innovation. Furthermore, it supports the notion that DI contributes to reducing CEI by fostering green innovation. Two plausible explanations can account for this phenomenon. Firstly, the integration of DI drives the advancement of green innovation, leading to the emergence of green production technologies that can be adopted to facilitate a transition toward environmentally friendly and low-carbon production practices and lifestyles. Secondly, DI facilitates the spillover of innovative green technologies, encourages regional cooperation, and amplifies the effects of CEI reduction.

4.4.5 *The Spatial Spillover Effects of DI*

It is noteworthy that DI has the ability to transcend spatial limitations, exerting influence not only on the local city but also extending its impact to surrounding cities. Several prior studies have highlighted the significant spatial dependence of urban carbon emissions (Zhang et al., 2023a, b). On the one hand, greenhouse gases can flow between different regions and exhibit pronounced spatial dependence (Li & Li, 2020). This means that if a city experiences high levels of carbon emissions in its atmosphere, the concentration of carbon emissions in neighboring cities is also likely to be high. Conversely, cities with low carbon dioxide concentrations often demonstrate aggregation effects. On the other hand, the development level of DI in local city also stimulates its construction in surrounding areas. Similarly, cities with well-developed DI are frequently surrounded by neighboring cities with similar characteristics (Liu et al., 2022).

More importantly, the influence of DI on urban CEI is not limited to local cities, but also influenced by neighboring cities. In particular, DI facilitates the exchange of environmental governance experience and technical cooperation among neighboring cities, due to its convenience and low cost (Rissman et al., 2020). Furthermore, DI expands the geographical flow boundaries of production factors, such as oil, coal, and water, and enhances resource utilization efficiency through remote monitoring, leading to a CEI reduction (Khan et al., 2018). Also, DI plays a role in spreading residents' environmental awareness and concepts to surrounding cities, thereby fostering regional environmental awareness. Consequently, this section employs a spatial econometric model to examine the spatial spillover effects of DI on urban CEI in China.

Spatial autocorrelation analysis The fundamental requirement for utilizing spatial econometrics to test spatial spillover effects is the presence of spatial autocorrelation. This chapter employs the Moran index to examine the spatial correlation and provides the following calculation formula:

$$I = \frac{\sum_{i=1}^{n}\sum_{j=1}^{n} w_{ij}\left(x_i - \bar{x}\right)\left(x_j - \bar{x}\right)}{S^2 \sum_{i=1}^{n}\sum_{j=1}^{n} w_{ij}} \tag{4.3}$$

where $\bar{x} = \frac{1}{n}\sum_{i=1}^{n} x_i$, $S^2 = \frac{1}{n}\sum_{i=1}^{n}\left(x_i - \bar{x}\right)^2$; x_i and x_j are the observed values in i and j city, respectively. w_{ij} is the element of the spatial weight matrix.

Furthermore, this chapter develops a neighboring spatial weight matrix that accurately depicts interrelationships among various cities to test the spatial impact of DI on CEI. The specific formula employed for this purpose is outlined below:

$$w_{ij} = \begin{cases} 1, & i \neq j \\ 0, & i = j \end{cases} \quad i, j = 1, 2, \cdots, n \tag{4.4}$$

where i and j are utilized to denote two distinct cities. A value of 1 is assigned when two cities are adjacent, while a value of 0 is assigned when they are not adjacent.

Table 4.8 displays the calculated results of the global Moran index for CEI. The findings reveal that the values of the Moran index are positive and their p-values for CEI in the years 2011–2019 are close to zero, suggesting a significant presence of positive spatial autocorrelation. This provides preliminary evidence of a notable spatial clustering among urban CEI in China. Additionally, it is observed that cities with high CEI tend to

Table 4.8 The global Moran index of CEI

Year	*I*	*E(I)*	*SD(I)*	*z*	*p-value*
2011	0.560	−0.004	0.040	14.059	0.000
2012	0.554	−0.004	0.040	13.903	0.000
2013	0.590	−0.004	0.040	14.806	0.000
2014	0.606	−0.004	0.040	15.212	0.000
2015	0.622	−0.004	0.040	15.601	0.000
2016	0.643	−0.004	0.040	16.125	0.000
2017	0.656	−0.004	0.040	16.481	0.000
2018	0.671	−0.004	0.040	16.839	0.000
2019	0.706	−0.004	0.040	17.694	0.000

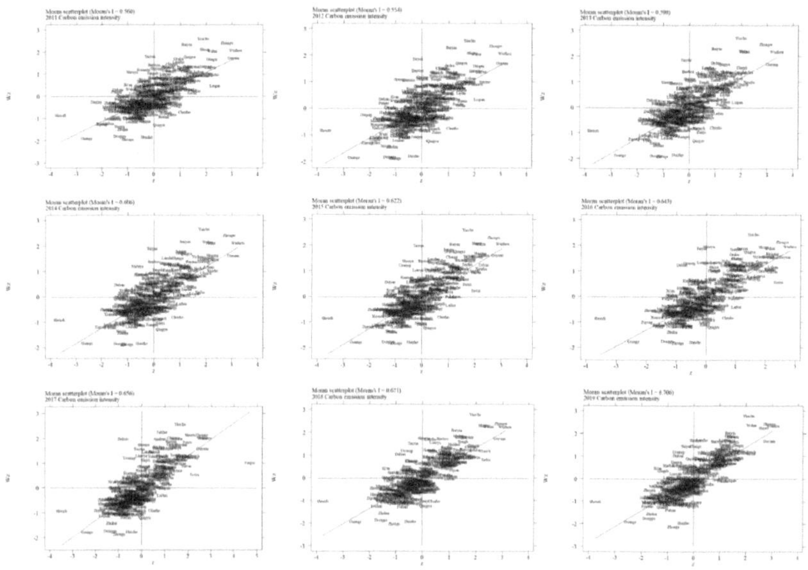

Fig. 4.1 Scatterplots of the local Moran index of CEI

be geographically adjacent to each other, further emphasizing this spatial clustering phenomenon

This chapter further examines the local Moran index to analyze the local spatial correlation between CEI. Figure 4.1 presents the scatterplots of the local Moran index of CEI spanning from 2011 to 2019. It is evident that CEI is predominantly concentrated in the first and third quadrants, indicating a positive spatial autocorrelation and the presence of value clustering.

Spatial analysis results This chapter adopts the spatial autoregression (SAR) model, spatial errors model (SEM), and spatial Durbin model (SDM), to explore the spatial spillover effects of DI on CEI. The formula is outlined as follows:

$$\begin{aligned} CEI_{it} &= \alpha_0 + \rho \mathrm{W} * CE_{it} + \beta DI_{it} + \gamma \mathrm{Control}_{it} + \sigma \mathrm{W} * X_{kit} + \theta_i + \omega_t + \mu_{it} \\ u_{it} &= \lambda W u_{it} + \varepsilon_{it}, \quad \varepsilon \sim N\left(0, \sigma^2 I_n\right) \end{aligned} \tag{4.5}$$

Table 4.9 Spatial regression results

Panel A: The estimation of spatial spillover effects			
Variables	SAR	SEM	SDM
	(1)	(2)	(3)
DI	−0.008*	−0.004	−0.006
	(0.004)	(0.004)	(0.004)
*W*DI*			−0.021***
			(0.008)
rho	0.496***		0.501***
	(0.000)		(0.000)
lambda		0.561***	
		(0.018)	
sigma2_e	0.008***	0.008***	0.008***
	(0.000)	(0.000)	(0.000)
Controls	Yes	Yes	Yes
City FE	Yes	Yes	Yes
Year FE	Yes	Yes	Yes
LR test (SAR)			29.600***
LR test (SEM)			93.460***
Observations	2466	2466	2466
R-squared	0.155	0.219	0.012
Panel B: Decomposition of the total marginal effects			
Variable	Direct effect	Indirect effect	Total effect
DI	−0.009**	−0.046***	−0.055***
	(0.004)	(0.014)	(0.017)

Robust standard errors in parentheses; *** $p < 0.01$, ** $p < 0.05$, * $p < 0.10$

where ρ represents the spatial lag regression coefficient, β is the regression coefficient of *DI*, γ is the regression coefficient of control variables, σ represents the spatial lag regression coefficient of *DI* and control variables. When $\rho \neq 0$ and $\lambda = \sigma = 0$, the model is SAR; when $\lambda \neq 0$ and $\rho = \sigma = 0$, the model is SEM; when $\rho \neq 0$, $\sigma \neq 0$, and $\lambda = 0$, the model is SDM; other variables are consistent with Eq. (4.2). In addition, during the spatial analysis, this chapter excludes samples of the revoked Zhangzhou city and the prefecture-level cities with missing data to construct a balanced panel dataset.

Table 4.9 Panel A presents the findings on spatial spillover effects using SAR, SEM, and SDM. The LR tests for SAR and SEM indicate significant at a 1% level, suggesting that SDM cannot be degraded into SAR and SEM. Following the methodology proposed by Elhorst (2014), this chapter determines that the SDM with city and year fixed effects is the optimal choice for conducting spatial effects analysis. Consequently, the discussion in this chapter focuses on the estimates presented in Column (3). The findings reveal that the spatial autoregressive coefficient ρ, determined by the neighborhood spatial weight matrix, is statistically remarkable at a 1% level. This signifies the presence of substantial spatial spillover effects in CEI between a local city and its neighboring cities. In other words, factors such as DI in both local and neighboring cities influence the local CEI. This finding aligns with the results reported by Zhang et al. (2023a, b). Additionally, the estimate of the spatial lag term of DI exhibits a remarkable negative association at a 1% level, thus providing initial evidence to support this relationship.

As SDM incorporates the simultaneous inclusion of spatial lags of both dependent and independent variables, it can generate a suite of spatial circular feedback effects that result in direct, indirect, and total effects (Elhorst, 2014). Consequently, it's worth noting that the estimates of DI do not equate to its marginal effects on CEI. To address this, this chapter decomposes the total marginal effects into direct, indirect and total effects, which are reported in Panel B of Table 4.9. The direct effect pertains to the overall impact of DI on CEI in the local city, while the indirect effect pertains to the overall impact of DI on CEI in neighboring cities, indicating a spatial spillover effect. Moreover, the total effect, which is the sum of the direct and indirect effects, quantifies the overall impact of DI on CEI in all cities.

The direct effect of DI is −0.009 and remarkable at a 5% level. This implies that for each standard deviation increase in DI, there is a reduction of 0.9% in CEI within the local city. The coefficient is comparatively lower than the benchmark results, primarily attributable to the spatial spillover effect of DI.

The indirect effect of DI is −0.046 and is also remarkable at a 1% level. This implies that DI has a spatial spillover effect on the CEI of surrounding cities. There are three main reasons for these findings. Firstly, carbon emissions can be diffused through natural factors like atmospheric circulation, or through economic mechanisms such as industrial transfer between cities (Li & Li, 2020). Consequently, the reduction in CEI resulting from DI construction in one city is replicated in neighboring cities. Secondly, DI reduces local market segmentation and decreases the cost of production factor circulation between cities. This reduction enhances the efficiency of urban resource allocation, leading to emission reduction (Zhang et al., 2023a, b)). Lastly, the technology spillover effect brought about by

DI construction also benefits neighboring areas, thus promoting the reduction of CEI in those areas.

4.5 Discussions

4.5.1 Contributions

This chapter makes three significant contributions. First, it expands our understanding of the relationship between DI and CEI. Previous research has predominantly focused on investigating the carbon effects of a particular aspect of DI, such as ICT, Internet technology, digital transportation, and broadband China policy. However, these studies have not comprehensively assessed the overall development of DI or its impact on carbon emissions. Consequently, they failed to accurately capture the variations in DI among cities and their corresponding effects on carbon emissions. One relevant study conducted by Tang and Yang (2023) used a number of keywords related to DI in government policy documents as an evaluation index for DI, examining its influence on urban carbon emissions in China. Interestingly, their findings contradicted the conclusions drawn in this chapter, suggesting a positive effect of DI on urban carbon emissions. However, accurately assessing the carbon emission impact of DI may be challenging due to policy delays. Therefore, this chapter develops multiple dimensions, including broadband Internet base, mobile Internet base, telecommunication base, information base, and software base, to construct a comprehensive DI index using the PCA. This index can be directly employed to measure the level of urban DI and its influence on the CEI of Chinese cities. Ultimately, this chapter conduces to a better understanding of the impact of DI on low-carbon development.

Second, this chapter highlights the effectiveness of DI in reducing urban CEI and offers insights into the regional variations and spatial spillover effects. Currently, there is a lack of reliable evidence to assess whether regional variations impact the relationship between DI and CEI. Furthermore, no studies have specifically addressed the spatial spillover effects between DI and CEI. Consequently, this chapter conduces to the existing literature by exploring the heterogeneous effects of DI on CEI across geographical divisions, specifically the north-south divided by Qinling–Huaihe Line and the east-west divided by Hu Huanyong Line, as well as regional locations and resource-based cities. Additionally, this chapter discusses the spatial effects of DI on CEI using a spatial Durbin model.

Third, this chapter contributes to the existing literature by examining the underlying mechanisms through which DI affects urban CEI. Previous research has predominantly concentrated on industrial structure and energy consumption, overlooking the significance of industrial upgrading, green innovation, and resource dependence. Consequently, this chapter aims to investigate the specific pathways by considering factors such as industrial upgrading, resource dependence, and green innovation.

4.5.2 Policy Implications

This chapter presents several practical implications based on prior research findings. Firstly, it is imperative for the government to prioritize the consolidation of DI in order to effectively reduce CEI. The current scale of DI construction is inadequate, leading to regional imbalances in its development. To address this issue, the government should concentrate on enhancing the level of DI development and systematically optimizing its configuration. Furthermore, a distinctive approach to DI development should be implemented, considering the first-mover advantage of DI in northern, central, and eastern cities, as well as in cities that have already established a mature and renewable resource base. Secondly, efforts should be made to narrow the regional gap in DI development and explore the energy-saving and carbon-reducing effects of DI in other cities. Thirdly, the government should leverage the catalytic role of DI in facilitating the upgrading of industrial structures, reducing reliance on traditional resources, and stimulating green innovation. Lastly, it is crucial to recognize the spatial spillover effects of DI on CEI. The reduction of CEI resulting from DI construction not only is confined to local cities but also impacts neighboring cities. Therefore, governments should enhance cooperation and exchange technical experiences in DI construction, which will lead to a doubled carbon reduction effect.

There are several limitations to this chapter. Firstly, due to a limited number of data samples, this chapter solely focuses on examining the influence of DI on the CEI for prefecture-level cities. Further exploration can expand to investigate the relationship between DI construction and CEI at the micro level, such as within enterprises and specific departments. Secondly, this chapter confirms that DI can reduce urban CEI by promoting green innovation. Future research should broaden its scope to explore the role of other factors, including cooperative innovation, independent innovation, incremental innovation, radical innovation, and innovation quality, in the link between DI and CEI. Thirdly, this chapter only discusses the spatial spillover effects of DI on the CEI for prefecture-level cities. Future research should expand its analysis to test whether the spillover effects of DI persist as the distance changes, and determine the maximum scope of this spillover effect.

4.6 Conclusions

China is presently undergoing a crucial phase of development and transformation, with the achievement of low-carbon transformation being a crucial component. DI plays a remarkable role in advancing the establishment of a digital China and serves as a momentous instrument to achieve the objectives of carbon peak and carbon neutrality objectives. Consequently, this chapter aims to investigate the impact of DI on CEI utilizing an unbalanced panel data of the prefecture-level cities from 2011 to 2019.

The findings of this chapter encompass several key aspects. Firstly, this chapter reveals that DI can significantly reduce urban CEI in China. This finding remains robust after performing various robustness tests and considering potential endogeneity issues. Secondly, the heterogeneity analysis reveals that the impact of DI on urban CEI is uneven and shows significant regional heterogeneity. Specifically, this effect is more pronounced in northern cities along the Qinling–Huaihe Line, cities located east of the Hu Huanyong Line, and cities in the central and eastern regions of Chinese regional divide. This regional disparity can be attributed to differences in industrial structure, population distribution, economic level, and scientific and technological strength. Thirdly, the findings confirm that the effect of DI on urban CEI is influenced by the type of resource-based city. Specifically, the reduction effect of DI on CEI is more prominent in resource-based cities, and further research indicates that this effect is particularly noticeable in resource-based cities at the stage of mature and regenerative development. These findings are closely related to the development mode and economic foundation of resource-based cities. Fourthly, the underlying mechanisms suggest that DI reduces urban CEI by promoting industrial upgrading, reducing resource dependence, and fostering green innovation. Lastly, this chapter analyzes the spatial spillover effects of DI, revealing a significant negative spillover effect on CEI. This implies that DI not only reduces CEI within a local city but also has a positive impact on CEI in surrounding cities.

References

Abraham, B. M. (2021). A subnational carbon curse? Fossil fuel richness and carbon intensity among US states. *The Extractive Industries and Society, 8*, 100859. https://doi.org/10.1016/j.exis.2020.12.007

Alam, M. M., & Murad, M. W. (2020). The impacts of economic growth, trade openness and technological progress on renewable energy use in organization for eco-nomic co-operation and development countries. *Renewable Energy, 145*, 382–390. https://doi.org/10.1016/j.renene.2019.06.054

Aldo, A. L., Andreas, B., & Martin, L. (2023). Exploring the associations between E-shopping and the share of shopping trip frequency and travelled time over total daily travel demand. *Travel Behaviour and Society, 31*, 202–208. https://doi.org/10.1016/j.tbs.2022.11.007

Ancillai, C., Sabatini, A., Gatti, M., & Perna, A. (2023). Digital technology and business model innovation: A systematic literature review and future research agenda. *Technological Forecasting and Social Change, 188*, 122307. https://doi.org/10.1016/j.techfore.2022.122307

Auty, R. (2002). *Sustaining development in mineral economies: The resource curse thesis*. Routledge.

Avotra, A. A. R. N., & Nawaz, A. (2023). Asymmetric impact of transportation on carbon emissions influencing SDGs of climate change. *Chemosphere, 324*, 138301. https://doi.org/10.1016/j.chemosphere.2023.138301

Bakken, B. E. (2019). Energy transition dynamics: Does participatory modelling contribute to alignment among differing future world views? *Systems Research and Behavioral Science, 36*, 184–196. https://doi.org/10.1002/sres.2578

Che, S., & Wang, J. (2022). Can environmental regulation solve the carbon curse of natural resource dependence: Evidence from China. *Resources Policy, 79*, 103100. https://doi.org/10.1016/j.resourpol.2022.103100

Chen, X., Shuai, C., Wu, Y., & Zhang, Y. (2020). Analysis on the carbon emission peaks of China's industrial, building, transport, and agricultural sectors. *Science of the Total Environment, 709*, 135768. https://doi.org/10.1016/j.scitotenv.2019.135768

Cheng, Z., Li, L., & Liu, J. (2018). Industrial structure, technical progress and carbon intensity in China's provinces. *Renewable & Sustainable Energy Reviews, 81*, 2935–2946. https://doi.org/10.1016/j.rser.2017.06.103

Colaco, R., & de Abreu e Silva, J. (2021). Exploring the interactions between online shopping, in-store shopping, and weekly travel behavior using a 7-day shopping survey in Lisbon, Portugal. *Transportation Research Record, 2675*, 379–390. https://doi.org/10.1177/0361198121990672

Dong, F., Yu, B., Hadachin, T., Dai, Y., Wang, Y., Zhang, S., & Long, R. (2018a). Drivers of carbon emission intensity change in China. *Resources Conservation and Recycling, 129*, 187–201. https://doi.org/10.1016/j.resconrec.2017.10.035

Dong, K., Sun, R., Hochman, G., & Li, H. (2018b). Energy intensity and energy conservation potential in China: A regional comparison perspective. *Energy, 155*, 782–795. https://doi.org/10.1016/j.energy.2018.05.053

Dong, F., Li, Y., Qin, C., Zhang, X., Chen, Y., Zhao, X., & Wang, C. (2022). Information infrastructure and greenhouse gas emission performance in urban China: A difference-in-differences analysis. *Journal of Environmental Management, 316*, 115252. https://doi.org/10.1016/j.jenvman.2022.115252

Elhorst, J. P. (2014). Matlab software for spatial panels. *International Regional Science Review, 37*, 389–405. https://doi.org/10.1177/0160017612452429

Fan, M., Li, M., Liu, J., & Shao, S. (2022). Is high natural resource dependence doomed to low carbon emission efficiency? Evidence from 283 cities in China. *Energy Economics, 115*, 106328. https://doi.org/10.1016/j.eneco.2022.106328

Fernández-Portillo, A., Almodóvar-González, M., & Hernández-Mogollón, R. (2020). Impact of ICT development on economic growth. A study of OECD European union countries. *Technology in Society, 63*, 101420. https://doi.org/10.1016/j.techsoc.2020.101420

Gao, Y., Lu, Y., Su, C. W., & Zhang, Y. (2023). Does China's low-carbon action reduce pollution emissions? A quasi-natural experiment based on the low-carbon city construction. *Environmental Science and Pollution Research, 30*, 27013–27029. https://doi.org/10.1007/s11356-022-24135-w

Gong, M., Zeng, Y., & Zhang, F. (2023). New infrastructure, optimization of resource allocation and upgrading of industrial structure. *Finance Research Letters, 54*, 103754. https://doi.org/10.1016/j.frl.2023.103754

Guo, Q., Wang, Y., & Dong, X. (2022). Effects of smart city construction on energy saving and CO_2 emission reduction: Evidence from China. *Applied Energy, 313*, 118879. https://doi.org/10.1016/j.apenergy.2022.118879

Haldar, A., & Sethi, N. (2022). Environmental effects of information and communication technology - Exploring the roles of renewable energy, innovation, trade and financial development. *Renewable and Sustainable Energy Reviews, 153*, 111754. https://doi.org/10.1016/j.rser.2021.111754

Hong, J., Shi, F., & Zheng, Y. (2023). Does network infrastructure construction reduce energy intensity? Based on the "broadband China" strategy. *Technological Forecasting and Social Change, 190*, 122437. https://doi.org/10.1016/j.techfore.2023.122437

Hu, J., Zhang, H., & Irfan, M. (2023). How does digital infrastructure construction affect low-carbon development? A multidimensional interpretation of evidence from China. *Journal of Cleaner Production, 396*, 136467. https://doi.org/10.1016/j.jclepro.2023.136467

Huang, Q., Yu, Y., & Zhang, S. (2019). Internet development and productivity growth in manufacturing industry: Internal mechanism and Chinese experiences. *China Industrial Economics, 8*, 5–23. In Chinese.

Huong, T. T. L., & Thanh, T. T. (2022). Is digitalization a driver to enhance environmental performance? An empirical investigation of European countries. *Sustainable Production and Consumption, 32*, 230–247. https://doi.org/10.1016/j.spc.2022.04.002

Hussain, J., & Lee, C. C. (2022). A green path towards sustainable development: Optimal behavior of the duopoly game model with carbon neutrality instruments. *Sustainable Development, 30*, 1523–1541. https://doi.org/10.1002/sd.2325

Jiang, Y., Zhao, X., & Zhai, L. (2023). Digital empowerment to improve the operational profitability in e-commerce supply chain. *Electronic Commerce Research and Applications, 58*, 101253. https://doi.org/10.1016/j.elerap.2023.101253

Khan, Z., Linares, P., Rutten, M., Parkinson, S., Johnson, N., & García-González, J. (2018). Spatial and temporal synchronization of water and energy systems: Towards a single integrated optimization model for long-term resource planning. *Applied Energy, 210*, 499–517. https://doi.org/10.1016/j.apenergy.2017.05.003

Lee, C. C., & Hussain, J. (2022). Optimal behavior of environmental regulations to reduce carbon emissions: A simulation-based dual green gaming model. *Environmental Science and Pollution Research, 29*, 56037–56054. https://doi.org/10.1007/s11356-022-19710-0

Lee, C. C., Chang, Y. F., & Wang, E. Z. (2022a). Crossing the rivers by feeling the stones: The effect of China's green credit policy on manufacturing firms' carbon emission intensity. *Energy Economics, 116*, 106413. https://doi.org/10.1016/j.eneco.2022.106413

Lee, C. C., Yuan, Y., & Wen, H. (2022b). Can digital economy alleviate CO_2 emissions in the transport sector? Evidence from provincial panel data in China. *Natural Resources Forum, 46*(3), 289–310. https://doi.org/10.1111/1477-8947.12258

Li, J., & Li, S. (2020). Energy investment, economic growth and carbon emissions in China—Empirical analysis based on spatial Durbin model. *Energy Policy, 140*, 111425. https://doi.org/10.1016/j.enpol.2020.111425

Li, Z., Dai, H., Song, J., Sun, L., Geng, Y., Lu, K., & Hanaoka, T. (2019). Assessment of the carbon emissions reduction potential of China's iron and steel industry based on a simulation analysis. *Energy, 183*, 279–290. https://doi.org/10.1016/j.energy.2019.06.099

Li, Y., Zhang, J., Yang, X., Wang, W., Wu, H., Ran, Q., & Luo, R. (2021). The impact of innovative city construction on ecological efficiency: A quasi-natural experiment from China. *Sustainable Production and Consumption, 28*, 1724–1735. https://doi.org/10.1016/j.spc.2021.09.012

Li, Y., Alharthi, M., Ahmad, I., Hanif, I., & Hassan, M. U. (2022). Nexus between renewable energy, natural resources and carbon emissions under the shadow of transboundary trade relationship from South East Asian economies. *Energy Strategy Reviews, 41*, 100855. https://doi.org/10.1016/j.esr.2022.100855

Li, M. J., Liang, S. F., Cheng, C., & Du, W. J. (2025). Digital infrastructure investment and carbon emissions reduction based on the Broadband China policy. *Utilities Policy, 95*, 101964. https://doi.org/10.1016/j.jup.2025.101964

Lin, B., & Wang, C. (2023). Does industrial relocation affect regional carbon intensity? Evidence from China's secondary industry. *Energy Policy, 173*, 113339. https://doi.org/10.1016/j.enpol.2022.113339

Liu, J., Yu, Q., Chen, Y., & Liu, J. (2022). The impact of digital technology development on carbon emissions: A spatial effect analysis for China. *Resources, Conservation and Recycling, 185*, 106445. https://doi.org/10.1016/j.resconrec.2022.106445

Lu, W. C. (2018). The impacts of information and communication technology, energy consumption, financial development, and economic growth on carbon dioxide emissions in 12 Asian countries. *Mitigation and Adaptation Strategies for Global Change, 23*, 1351–1365. https://doi.org/10.1007/s11027-018-9787-y

Madlener, R., Sheykhha, S., & Briglauer, W. (2022). The electricity-and CO_2-saving potentials offered by regulation of European video-streaming services. *Energy Policy, 161*, 112716. https://doi.org/10.1016/j.enpol.2021.112716

Martínez, J. M. G., Puertas, R., Martín, J. M. M., & Ribeiro-Soriano, D. (2022). Digitalization, innovation and environmental policies aimed at achieving sustainable production. *Sustainable Production and Consumption, 32*, 92–100. https://doi.org/10.1016/j.spc.2022.03.035

Nudurupati, S. S., Budhwar, P., Pappu, R. P., Chowdhury, S., Kondala, M., Chakraborty, A., & Ghosh, S. K. (2022). Transforming sustainability of Indian small and medium-sized enterprises through circular economy adoption. *Journal of Business Research, 149*, 250–269. https://doi.org/10.1016/j.jbusres.2022.05.036

Raheem, I. D., Tiwari, A. K., & Balsalobre-Lorente, D. (2020). The role of ICT and financial development in CO_2 emissions and economic growth. *Environmental Science and Pollution Research, 27*, 1912–1922. https://doi.org/10.1007/s11356-019-06590-0

Ren, S., Hao, Y., Xu, L., Wu, H., & Ba, N. (2021). Digitalization and energy: How does internet development affect China's energy consumption? *Energy Economics, 98*, 105220. https://doi.org/10.1016/j.eneco.2021.105220

Ren, X., Zeng, G., & Gozgor, G. (2023). How does digital finance affect industrial structure upgrading? Evidence from Chinese prefecture-level cities. *Journal of Environmental Management, 330*, 117125. https://doi.org/10.1016/j.jenvman.2022.117125

Rissman, J., Bataille, C., Masanet, E., Aden, N., Morrow, W. R., Zhou, N., Elliott, N., Dell, R., Heeren, N., & Huckestein, B. (2020). Technologies and policies to decarbonize global industry: Review and assessment of mitigation drivers through 2070. *Applied Energy, 266*, 114848. https://doi.org/10.1016/j.apenergy.2020.114848

Rukanova, B., van Engelenburg, S., Ubacht, J., Tan, Y. H., Geurts, M., Sies, M., Molenhuis, M., Slegt, M., & van Dijk, D. (2023). Public value creation through voluntary business to government information sharing enabled by digital infrastructure innovations: A framework for analysis. *Government Information Quarterly, 40*, 101786. https://doi.org/10.1016/j.giq.2022.101786

Su, C. W., Pang, L. D., Tao, R., Shao, X., & Umar, M. (2022a). Renewable energy and technological innovation: Which one is the winner in promoting net-zero emissions? *Technological Forecasting and Social Change, 182*, 121798. https://doi.org/10.1016/j.techfore.2022.121798

Su, C. W., Umar, M., & Gao, R. (2022b). Save the environment, get financing! How China is protecting the environment with green credit policies? *Journal of Environmental Management, 323*, 116178. https://doi.org/10.1016/j.jenvman.2022.116178

Su, C. W., Liu, F., Stefea, P., & Umar, M. (2023). Does technology innovation help to achieve carbon neutrality? *Economic Analysis and Policy, 78*, 1–14. https://doi.org/10.1016/j.eap.2023.01.010

Tang, K., & Yang, G. (2023). Does digital infrastructure cut carbon emissions in Chinese cities? *Sustainable Production and Consumption, 35*, 431–443. https://doi.org/10.1016/j.spc.2022.11.022

Tao, W. L., Weng, S. M., Chen, X. L., & Song, M. L. (2025). Breaking the inertia of traditional economic development: Does network infrastructure construction achieve urban carbon unlocking? *Sustainable Cities and Society, 121*, 106197. https://doi.org/10.1016/j.scs.2025.106197

Wang, K., & Deng, Y. (2016). Can new urbanization break through the Hu Huanyong Line? Further discussion on the geographical connotations of the Hu Huanyong Line. *Geographical Research, 35*, 825–835. In Chinese.

Wang, L., & Shao, J. (2023). Digital economy, entrepreneurship and energy efficiency. *Energy, 269*, 126801. https://doi.org/10.1016/j.energy.2023.126801

Wang, L. H., & Shao, J. (2024). The energy saving effects of digital infrastructure construction: Empirical evidence from Chinese industry. *Energy, 294*, 130778. https://doi.org/10.1016/j.energy.2024.130778

Wang, K., Wu, M., Sun, Y., Shi, X., Sun, A., & Zhang, P. (2019). Resource abundance, industrial structure, and regional carbon emissions efficiency in China. *Resources Policy, 60*, 203–214. https://doi.org/10.1016/j.resourpol.2019.01.001

Wang, Q., Hu, A., & Tian, Z. (2022). Digital transformation and electricity consumption: Evidence from the Broadband China pilot policy. *Energy Economics, 115*, 106346. https://doi.org/10.1016/j.eneco.2022.106346

Wang, H., Li, Y., Lin, W., & Wei, W. (2023). How does digital technology promote carbon emission reduction? Empirical evidence based on e-commerce pilot city policy in China. *Journal of Environmental Management, 325*, 116524. https://doi.org/10.1016/j.jenvman.2022.116524

Wu, L., Sun, L., Qi, P., Ren, X., & Sun, X. (2021). Energy endowment, industrial structure upgrading, and CO_2 emissions in China: Revisiting resource curse in the context of carbon emissions. *Resources Policy, 74*, 102329. https://doi.org/10.1016/j.resourpol.2021.102329

Xu, X., Watts, A., & Reed, M. (2019). Does access to internet promote innovation? A look at the US broadband industry. *Growth and Change, 50*, 1423–1440. https://doi.org/10.1111/grow.12334

Yahya, F., & Lee, C. C. (2023). The asymmetric effect of agriculturalization toward climate neutrality targets. *Journal of Environmental Management, 328*, 116995. https://doi.org/10.1016/j.jenvman.2022.116995

Yang, G., Li, D., & Wen, Y. (2022a). Does network infrastructure improve the information efficiency of regional capital market?—Quasi natural experiment based on "Broadband China" strategy, *International Conference on Business Intelligence and Information Technology.*

Yang, Z., Gao, W., Han, Q., Qi, L., Cui, Y., & Chen, Y. (2022b). Digitalization and carbon emissions: How does digital city construction affect China's carbon emission reduction? *Sustainable Cities and Society, 87*, 104201. https://doi.org/10.1016/j.scs.2022.104201

Yao, L., Li, A. Y., & Yan, E. W. (2025). Research on digital infrastructure construction empowering new quality productivity. *Scientific Reports, 15*(1), 6645. https://doi.org/10.1038/s41598-025-90811-9

Yi, M., Liu, Y., Sheng, M. S., & Wen, L. (2022). Effects of digital economy on carbon emission reduction: New evidence from China. *Energy Policy, 171*, 113271. https://doi.org/10.1016/j.enpol.2022.113271

Yi, M., Chen, D. H., Wu, T., Tao, M. M., Sheng, M. S., & Zhang, Y. (2024). Intelligence and carbon emissions: The impact of smart infrastructure on carbon emission intensity in cities of China. *Sustainable Cities and Society, 112*, 105602. https://doi.org/10.1016/j.scs.2024.105602

You, J., & Zhang, W. (2022). How heterogeneous technological progress promotes industrial structure upgrading and industrial carbon efficiency? Evidence from China's industries. *Energy, 247*, 123386. https://doi.org/10.1016/j.energy.2022.123386

Yu, Z. X., Liu, S., & Li, S. Q. (2024). Research on the spatial effect of digital economy development on urban carbon reduction. *Journal of Environmental Management, 357*, 120764. https://doi.org/10.1016/j.jenvman.2024.120764

Yu, T. H., Zhang, Y., Jia, S. S., & Cui, X. F. (2025). Spatio-temporal evolution and drivers of coupling coordination between digital infrastructure and inclusive green growth: Evidence from the Yangtze River economic belt. *Journal of Environmental Management, 376*, 124416. https://doi.org/10.1016/j.jenvman.2025.124416

Zeng, M., & Zhang, W. K. (2024). Green finance: The catalyst for artificial intelligence and energy efficiency in Chinese urban sustainable development. *Energy Economics, 139*, 107883. https://doi.org/10.1016/j.eneco.2024.107883

Zhang, M., & Liu, Y. (2022). Influence of digital finance and green technology innovation on China's carbon emission efficiency: Empirical analysis based on spatial metrology. *Science of the Total Environment, 838*, 156463. https://doi.org/10.1016/j.scitotenv.2022.156463

Zhang, W. K., & Zeng, M. (2024). Is artificial intelligence a curse or a blessing for enterprise energy intensity? Evidence from China. *Energy Economics, 134*, 107561. https://doi.org/10.1016/j.eneco.2024.107561

Zhang, L., Mu, R., Zhan, Y., Yu, J., Liu, L., Yu, Y., & Zhang, J. (2022a). Digital economy, energy efficiency, and carbon emissions: Evidence from provincial panel data in China. *Science of the Total Environment, 852*, 158403. https://doi.org/10.1016/j.scitotenv.2022.158403

Zhang, W., Liu, X., Wang, D., & Zhou, J. (2022b). Digital economy and carbon emission performance: Evidence at China's city level. *Energy Policy, 165*, 112927. https://doi.org/10.1016/j.enpol.2022.112927

Zhang, W. K., Luo, Q., & Liu, S. (2022c). Is government regulation a push for corporate environmental performance? Evidence from China. *Economic Analysis and Policy, 74*, 105–121. https://doi.org/10.1016/j.eap.2022.01.018

Zhang, W. K., Fan, H. X., & Zhao, Q. W. (2023a). Seeing green: How does digital infrastructure affect carbon emission intensity? *Energy Economics, 127*, 107085. https://doi.org/10.1016/j.eneco.2023.107085

Zhang, W. K., Zeng, M., Zhang, Y., & Su, C. W. (2023b). Reducing carbon emissions: Can high-speed railway contribute? *Journal of Cleaner Production, 413*, 137524. https://doi.org/10.1016/j.jclepro.2023.137524

Zhang, W. K., Fan, H. X., & Zeng, M. (2025). Unleashing the power of artificial intelligence: A game changer for urban energy efficiency in China. *Sustainable Cities and Society, 126*, 106372. https://doi.org/10.1016/j.scs.2025.106372

Zhao, M., & Sun, T. (2022). Dynamic spatial spillover effect of new energy vehicle industry policies on carbon emission of transportation sector in China. *Energy Policy, 165*, 112991. https://doi.org/10.1016/j.enpol.2022.112991

Zhao, Y., Zhao, Z., Qian, Z., Zheng, L., Fan, S., & Zuo, S. (2023). Is cooperative green innovation better for carbon reduction? Evidence from China. *Journal of Cleaner Production, 394*, 136400. https://doi.org/10.1016/j.jclepro.2023.136400

Zheng, S., Ahmed, D., Xie, Y., Majeed, M. T., & Hafeez, M. (2023). Green growth and carbon neutrality targets in China: Do financial integration and ICT matter? *Journal of Cleaner Production, 405*, 136923. https://doi.org/10.1016/j.jclepro.2023.136923

PART II

Enterprise-Level Analysis

CHAPTER 5

AI-Enterprise ESG Synergy

5.1 Introduction

The increasing challenges associated with global environmental degradation (Su et al., 2022; Zeng & Zhang, 2024; Zhang & Zeng, 2024; Zhang et al., 2025), a lack of social responsibility, and persistent governance scandals (Su et al., 2023a) have rendered the conventional enterprise evaluation framework—primarily focused on financial performance—insufficient for meeting the requirements of sustainable development (Zhu et al., 2023). In this regard, the environmental, social, and governance (ESG) framework, initially introduced by the United Nations Global Compact in 2004, has rapidly gained prominence as a critical paradigm for the strategic transformation of enterprises globally (Vezeteu et al., 2024). Over the years, the implementation of ESG practices within enterprises has become integral to sustainable development (Wan et al., 2024). According to the Global Sustainable Investment Alliance (GSIA),[1] the global allocation for ESG investments reaches $5.588 trillion in 2022, representing over

[1] GSIA: https://www.gsi-alliance.org/members-resources/gsir2022/

The core content of this chapter is derived from this article: Zhang, W. K., & Zeng, M. (2025). When AI Powers Sustainability: Divergent Paths for Enterprise ESG Performance in China. *Romanian Journal of Economic Forecasting, 28 (3)*, 24–43.

W. Zhang et al., *Green Transformation in the Digital Age*,
https://doi.org/10.1007/978-981-95-9121-3_5

one-sixth of the total sustainable investment market. Enhancing the preparedness and commitment of enterprises to adopt ESG practices—termed "enterprise ESG performance"—is essential for advancing ESG initiatives and promoting sustainable development (Guo et al., 2024a). However, the implementation of ESG practices often incurs significant costs and generates positive externalities, which can substantially hinder enterprise ESG performance (Gao & Liu, 2023). According to the Institute of Management Accountants (IMA),[2] merely 33% of Chinese sample enterprises were involved in ESG practices in the year 2022. Hence, improving enterprise ESG performance has become prominent topics in contemporary academic discussions (Gao et al., 2024).

Artificial intelligence (AI) can greatly affect enterprises (Zhang & Zeng, 2024). It has the potential to improve operational efficiency and increase profitability, thereby providing a solid material basis for enterprises to engage in ESG practices (Chen et al., 2024; Yu, 2024). Additionally, AI can enhance the visibility and influence of enterprises, amplify the anticipated advantages associated with ESG initiatives, and address the positive externalities inherent in ESG, thus motivating enterprises to adopt these practices (Chen & Zhang, 2024). It is clear that AI offers a promising avenue for improving enterprise ESG performance. However, their relationship remains inadequately examined (Huang et al., 2024). There is a lack of comprehensive analysis regarding how AI improves enterprise ESG performance, and empirical evidence to support such claims is limited.

The relevant literature for this chapter includes: the AI's impact and the determinants influencing enterprise ESG performance. The former emphasizes its economic ramifications (Qin et al., 2024a), such as fostering economic growth (Deng et al., 2023), driving innovation (Babina et al., 2024), and increasing resource efficiency (Zhang & Zeng, 2024). However, limited research explores its noneconomic implications, particularly its influence on ESG performance. While substantial scholarship exists regarding AI's contributions to environmental protection (the E item) (Zhang et al., 2025), few studies have explored social responsibility (the S item) and enterprise governance (the G item) (Chen et al., 2024).

Conversely, the literature concerning enterprise ESG performance tends to concentrate on enterprise-specific attributes, including technical emphasis (Fang et al., 2023), management competence (Fan et al., 2024), and enterprise culture (Bai et al., 2024). Nonetheless, limited research directly examines AI-enterprise ESG performance nexus. Notably, Chen et al. (2024) employ text analysis techniques to assess the extent of AI

[2] IMA: https://www.imachina.org.cn/Uploads/File/2023/08/02/u64c9cd95905eb.pdf

utilization within enterprises, demonstrating that a broader application of AI correlates with enhanced ESG performance. Similarly, Li et al. (2024a) investigate the AI adoption–ESG performance nexus, concluding a positive association. Furthermore, Lin and Zhu (2025) and Huang et al. (2024), utilizing China's AI pilot policy, illustrate that AI can improve enterprise ESG performance.

While these studies have contributed to understanding the AI-enterprise ESG performance nexus, they fail to delve into the theoretical foundations of this nexus at the city level. Undertaking research at the city level can significantly improve the validity of econometric regression outcomes. Furthermore, it facilitates a more comprehensive investigation of associated subjects, such as the examination of the heterogeneity among various types of cities. Additionally, they do not address the underlying mechanism of analyst attention and digital transformation, nor do they consider the regulatory effects of marketization level and business environment. In light of this context, this chapter is conducted. The analysis is based on a dataset that includes 1270 publicly listed companies in China, spanning 187 cities from 2006 to 2019. Initially, we employ a city–year fixed effects model to establish that AI positively influences enterprise ESG performance. And it remains valid through various robustness tests, including alterations to the dependent and independent variables as well as different regression methodologies. Subsequently, we carry out the heterogeneity analysis through the lenses of ESG items, city classifications, industry categories, and types of enterprises. Finally, we investigate the underlying mechanisms through which AI affects ESG performance, specifically examining whether AI enhances enterprise ESG performance via increased analyst attention and digital transformation, and whether marketization level and business environment amplify AI's effectiveness in improving enterprise ESG performance.

5.2 Theoretical Analysis

5.2.1 Impact of AI on Enterprise ESG Performance

The current literature suggests that how AI affects enterprise ESG performance can be analyzed through capability and motivation. From the capability perspective, AI significantly bolsters enterprises' ability to protect the environment (Zhang & Zeng, 2024). It promotes innovative activities and technological advancements, thereby enabling enterprises to adopt cutting-edge green production technologies and efficient manufacturing processes (Li et al., 2023). Consequently, this helps the carbon and pollutant emission reduction (Zhang et al., 2023a). Additionally, AI plays a crucial role in optimizing resource utilization and minimizing waste (Song et al., 2024). For example, it aids enterprises in selecting low-carbon, high-quality supply chains (Qin et al., 2024b) and enhances real-time energy monitoring and optimization of energy usage strategies.

Furthermore, AI improves production processes by leveraging modern digital technologies, which increases labor efficiency and overall production capacity (Chen et al., 2024). This enhancement is advantageous for boosting business revenue, thereby creating a foundation for enterprises to meet their social responsibilities (Chen et al., 2024). Lastly, AI assists enterprises in identifying, analyzing, and managing various operational risks through the application of advanced models developed using deep learning techniques (Lee et al., 2024; Yang & Wu, 2022). This utilization of AI reduces uncertainty in enterprise management, enhances the effectiveness of risk control measures, and ultimately contributes to improved governance outcomes for enterprises.

From a motivational perspective, AI enhances the transparency of enterprise environmental protection information, thereby reducing market information asymmetry and increasing enterprise engagement in environmental initiatives (Wang et al., 2025). By strengthening societal oversight of enterprise environmental practices, AI promotes the internalization of external costs associated with environmental pollution, thereby compelling enterprises to reduce emissions and invest more in environmental governance (Huo & Wang, 2023). Additionally, AI amplifies the societal impact of enterprise participation in environmental protection, assisting companies in cultivating a responsible enterprise image and securing financing opportunities, which further incentivizes them to engage in additional environmental protection activities (Chen & Zhang, 2024).

Furthermore, AI can aid enterprises in developing a comprehensive language model that assesses the emotional sentiments expressed on social media by both employees and consumers (Wang et al., 2022). This capability enhances communication between enterprises and their stakeholders, facilitating a deeper understanding of stakeholder expectations. Concurrently, AI encourages enterprises to adapt their policies promptly in response to these expectations, thereby effectively fulfilling their social responsibilities. Lastly, AI can support directors and senior management in fostering an awareness of enterprise governance, enhancing their understanding of governance principles, and guiding the establishment of a modern enterprise management framework (Johnson, 2019), which includes the reinforcement of internal controls, the development of a fair compensation structure, and the equitable distribution of authority within the organization.

Therefore, this chapter proposes the following hypothesis:

Hypothesis 5.1: AI can significantly facilitate enterprise ESG performance.

5.2.2 Underlying Mechanism of AI Affects Enterprise ESG Performance

This chapter investigates the underlying mechanism of AI affects enterprise ESG performance from the perspectives of both pressure and power. From the pressure standpoint, AI may influence enterprise ESG performance by shaping analyst attention. Additionally, from the power standpoint, AI can affect enterprise ESG performance by facilitating digital transformation.

Analyst attention Analyst attention refers to the mechanism by which analysts convey the operational status of enterprises to the capital market through the examination of both publicly accessible and confidential information. This attention is pivotal in improving enterprise ESG performance (Wang & Cao, 2024). Firstly, analyst attention bridges the gap between enterprises, investors, and consumers, thereby alleviating information asymmetry (Huang et al., 2025). Increased transparency in the dissemination of information exerts external pressure on enterprises to prioritize ESG initiatives, as both investors and consumers are progressively emphasizing the importance of sustainability practices and are more likely to engage with enterprises that exhibit social and environmental responsibility. Secondly, analyst attention facilitates the public disclosure of enterprises' information and practices, which can motivate enterprises to invest in ESG initiatives (Li et al., 2024b). In an effort to cultivate a responsible enterprise image and enhance their competitive position in the market, enterprises are increasingly engaging in ESG-related activities.

AI has the potential to augment analyst attention toward enterprises, thereby enhancing their ESG performance (Imjai et al., 2025). Firstly, modern digital technological infrastructures, such as 5G networks and cloud computing—integral components of AI—provide a robust framework for analysts to collect and analyze extensive, real-time data related to enterprises. Secondly, AI equips analysts with sophisticated analytical tools that facilitate a scientific and precise assessment of business operations. Finally, AI streamlines the report-writing process for analysts by automating the generation of financial evaluation reports and comprehensive analytical recommendations, thereby improving their overall productivity.

Therefore, this chapter proposes the following hypothesis:

Hypothesis 5.2: AI can facilitate enterprise ESG performance by enhancing analyst attention.

Digital transformation Digital transformation encompasses the comprehensive integration of digital technologies, leading to fundamental changes in production processes, management systems, and business models (Cai et al., 2023). The introduction of advanced and efficient digital technologies is crucial for optimizing production chains, enhancing resource efficiency, and reducing pollution emissions (Su et al., 2023a, b; Zhang et al., 2024). Additionally, digital transformation promotes a more streamlined, transparent, and hierarchical management structure within enterprises, effectively aligning various stakeholders' interests. This alignment strengthens the implementation of enterprise social responsibility initiatives and raises the standards of internal governance (Cai et al., 2023).

The relationship between AI and enterprise digital transformation is characterized by technological isomorphism, which serves as a foundational technical basis for organizations pursuing digital transformation initiatives (Holmstrom, 2022). AI not only accelerates the adoption and advancement of digital technologies within enterprises but also aids in the integration and analysis of heterogeneous data from various sources, thereby contributing to the accumulation of data assets. Moreover, AI enables enterprises to automate production processes, replacing repetitive labor, modernizing production workflows and business models, and advancing the overall digital transformation of enterprises (Taherizadeh & Beaudry, 2023).

Therefore, this chapter proposes the following hypothesis:

Hypothesis 5.3: AI can facilitate enterprise ESG performance by stimulating digital transformation.

5.2.3 Regulatory Effects of AI on Enterprise ESG Performance

Marketization level Marketization level is conceptualized as the degree to which a market economy forms the foundational economic structure of a region, serving as a metric for economic maturity and the marketization process (Wang et al., 2021). A higher level of marketization is correlated with a more liberal movement of production factors, a more robust framework for the protection of property rights, and increased market competition. Theoretically, marketization level enhances the impact of AI on enterprise ESG performance.

Firstly, regions with higher levels of marketization are more proficient in attracting and retaining digital talent, technology, data, and other essential resources, which in turn fosters improvements in AI capabilities and supports enterprise ESG initiatives (Fu et al., 2025). Secondly, a significant degree of marketization reflects diminished governmental intervention and heightened market competition, which compels enterprise leaders to embrace sustainable ESG principles and utilize AI technologies to bolster their ESG efforts (An et al., 2023). Finally, in regions characterized by advanced marketization, the regulatory frameworks governing market practices, including ESG information disclosure, are typically more comprehensive and standardized, thereby incentivizing enterprises to prioritize their ESG practices (You et al., 2025).

Therefore, this chapter proposes the following hypothesis:

Hypothesis 5.4: Marketization level can enhance the positive impact of AI on enterprise ESG performance.

Business environment Business environment refers to the institutional frameworks that enterprises encounter in their operational activities, encompassing elements such as governmental efficiency, the legal system, the financial landscape, and the degree of international integration (Luo et al., 2023). From a theoretical perspective, a conducive business environment enhances the capacity of AI to improve enterprise ESG performance.

Firstly, a favorable business environment enables enterprises to take advantage of efficient administrative processes and substantial tax incentives (Zhang et al., 2023b), which in turn lowers operational costs and creates a conducive foundation for AI to effectively enhance ESG performance. Secondly, a strong innovation protection system, which is a hallmark of a favorable business environment, motivates enterprises to increase their adoption of AI technologies and their willingness to invest in ESG initiatives (Luo et al., 2023). Furthermore, a supportive business environment enhances access to financing, thereby providing critical financial resources necessary for enterprises to execute ESG activities (Guo et al., 2024b). Finally, a favorable business environment is associated with a greater alignment between domestic institutional regulations in China and international standards. This alignment has facilitated the development of systematic and standardized guidelines for enterprise ESG information disclosure on a global scale, which can further encourage Chinese enterprises to engage in ESG practices.

Therefore, this chapter proposes the following hypothesis:

Hypothesis 5.5: Business environment can enhance the positive impact of AI on enterprise ESG performance.

5.3 Research Design

5.3.1 Data

The AI-related data is from the International Federation of Robotics (IFR).[3] It offers data regarding the annual installation of new industrial robots, as well as the number of operational industrial robots across diverse industries in China. This information can be further disaggregated to the city level, thereby serving as an indicator of the AI capabilities present in Chinese cities. Given that IFR currently provides data only up to the year 2019, the sample period ends in 2019. The ESG-related data is from the Bloomberg.[4] And the other data is from the China Stock Market and Accounting Research Database (CSMAR).[5] Following the removal of samples containing missing values and outliers, this chapter successfully compiled a dataset comprising 1270 enterprises across 187 cities, resulting in a total of 11,737 observations spanning from 2006 to 2019.

[3] IFR: https://ifr.org/
[4] Bloomberg: https://www.bloomberg.com/
[5] CSMAR: https://data.csmar.com/

5.3.2 Variable

The explained variable Major agencies evaluated Chinese enterprises' ESG performance, including Bloomberg, Sino-Securities Index, SynTao Green Finance, Wind, and others. Notably, Bloomberg has emerged as a leader in the field of ESG rating, having developed a comprehensive and systematic index that encompasses 46 indicators related to environmental dimensions, 46 indicators pertaining to social dimensions, and 30 indicators associated with governance dimensions (Chen & Xie, 2022). Consequently, Bloomberg's ESG data is extensively utilized within academic research, owing to its substantial sample size and rigorous evaluation framework, such as Khurram et al. (2024), Liu et al. (2024), and Nollet et al. (2016). Hence, we introduce the Bloomberg ESG score as a metric to assess Chinese enterprises' ESG performance. Simultaneously, we employ the Sino-Securities Index as an alternative proxy variable to assess Chinese enterprises' ESG performance, thereby conducting robustness checks to strengthen the validity of the research findings.

The explanatory variable AI encompasses a wide array of concepts, which complicates the systematic assessment of AI levels across Chinese cities. In this regard, current research frequently employs industrial robot data to measure AI levels (Zeng & Zhang, 2024). An industrial robot is defined as an automated intelligent tool designed to replicate human actions within the industrial production process. These robots exhibit key characteristics associated with AI, namely intelligence, autonomy, and adaptability, thereby positioning them as a central focus in the advancement of AI technologies (Acemoglu & Restrepo, 2020). Following Zeng & Zhang (2024), this chapter measures AI by the ratio of urban industrial robots to their employment levels. The specific formula is shown in Eq. (5.1):

$$\mathrm{AI}_{it} = \sum_{k=1}^{K} \frac{\mathrm{empl}_{i,k,t=2005}}{\mathrm{empl}_{i,t=2005}} \times \frac{\mathrm{robot}_{kt}}{\mathrm{empl}_{k,t=2005}} \tag{5.1}$$

where i, k, and t represent city, industry, and year, respectively; *robot* denotes number of industrial robots, and empl indicates employment.

This calculation methodology fundamentally incorporates the primary concept of Bartik instrumental variable method, thereby serving as an effective indicator of urban AI level. Besides, this chapter measures AI

Table 5.1 Descriptive statistics

Variables	*N*	*Mean*	*S.D.*	*Min*	*Max*
Dependent variable					
ESG	11,737	25.237	8.692	6.198	65.778
Independent variable					
AI_install (units/hundred workers)	11,737	0.069	0.060	0.002	0.285
AI_stock (units/hundred workers)	11,737	0.281	0.270	0.007	1.424
Controls					
The enterprises' annual net profit (billion RMB)	11,737	2.366	13.744	−17.049	313.361
State-owned enterprise or not (yes=1; no=0)	11,737	0.553	0.497	0	1
The ratio of liabilities to assets	11,737	0.503	0.255	0.008	13.629
The shares of the top five shareholders (0,1)	11,737	0.546	0.209	0.200	0.991
Tobin's Q value	11,737	2.560	2.329	0.205	45.651
The number of directors with overseas experience (unit)	11,737	0.558	1.015	0	9
The number of independent directors (unit)	11,737	3.427	0.796	0	8

based on newly installed robots (AI_install) and existing robot stock (AI_stock).

The control variables This chapter introduces control variables, as presented in Table 5.1.

Table 5.1 provides these variables' descriptive statistics. As shown, the average value of ESG is 25.237. It agrees with the value 20.35 reported by Chen and Xie (2022), who also employ Bloomberg ESG data to assess the ESG performance of Chinese enterprises. Consequently, it can be inferred that the data utilized in this research, following a thorough cleaning process, is dependable.

5.3.3 Method

This chapter establishes the following model to explore how AI affects Chinese enterprise' ESG performance.

$$ESG_{jt} = \alpha_0 + \beta \mathrm{AI}_{jit} + \eta C_{jit} + u_j + v_t + \varepsilon_{jt} \tag{5.2}$$

where j, i, and t represent enterprise, city, and year, respectively; ESG is enterprise ESG performance; AI is urban AI level, and β captures how AI affects ESG; C represents these control variables; u_j and v_t are the enterprise and year fixed-effects (FE) respectively; α_0 is the intercepted item, and ε_{jt} is a random error vector.

5.4 Findings

5.4.1 *Baseline Results*

Table 5.2 provides the results of Eq. (5.2). All columns exhibited statistically significant positive regression coefficients. The findings indicate that the implementation of AI enhances enterprise ESG performance. In particular, the coefficient associated with AI_install in Column (2) is 7.855. This indicates that a 1 standard deviation increase in the number of newly installed industrial robots per hundred workers within Chinese cities is associated with a 0.054 (7.855 × 0.060/8.692) standard deviation increase in the ESG performance score of local enterprises. Similarly, 1.792 for AI_stock in Column (4) means that each extra standard deviation in the number of urban industrial robots per hundred workers enhances enterprises' ESG performance by 0.056 (1.792 × 0.270/8.692) standard deviations. In conclusion, the benchmark regression analysis substantiates the assertion that AI positively influences enterprise ESG performance, thereby corroborating Hypothesis 5.1.

Table 5.2 Baseline regression results

	(1)	*(2)*	*(3)*	*(4)*
AI_install	8.407***	7.855***		
	(2.586)	(2.582)		
AI_stock			1.871***	1.792***
			(0.564)	(0.567)
Controls	No	Yes	No	Yes
Enterprise fixed-effect	Yes	Yes	Yes	Yes
Year fixed-effect	Yes	Yes	Yes	Yes
N	11,737	11,737	11,737	11,737
R^2	0.842	0.843	0.842	0.843

*** $p < 0.01$. Robust standard errors are in parentheses

The conclusions drawn in this chapter are fundamentally aligned with existing literature in the field. Chen et al. (2024) utilized text analysis to assess the extent of AI implementation within enterprises, establishing that a higher level of AI integration correlates with improved ESG performance. Furthermore, Chen and Zhang (2025) conducted a multivariate regression analysis, demonstrating that AI applications within enterprises facilitate enhanced ESG performance. Similarly, Li et al. (2024a, b) identified their significant positive relationship. Additionally, Huang et al. (2024) and Lin and Zhu (2025) applied the difference-in-differences (DID) methodology to evidence that the AI policies improve enterprises' ESG performance. While the perspectives of these studies differ from those presented in this chapter, their findings are congruent with the conclusions of this chapter. This alignment among various studies serves to reinforce the validity of the findings presented herein.

5.4.2 Robustness Tests

Alternative measure for enterprise ESG performance In numerous organizations focused on enterprise ESG assessments, the Sino-Securities Index plays a significant role and has gained considerable traction within academic discourse (Kuai et al., 2025). This index is developed by Chinese ESG evaluation organizations with the help of the algorithm-based big data method. Following Kuai et al. (2025), this chapter reevaluates enterprise ESG performance by utilizing the Sino-Securities Index. The results in Table 5.3 indicate that the positive coefficients for AI_install and AI_stock further substantiate the baseline assertion.

Alternative measure for AI Drawing inspiration from Zeng and Zhang (2024)'s methodology, this chapter employs the AI-enterprise density to reevaluate Chinese urban AI level, quantified by the number of AI enterprises per thousand residents. Enterprises that incorporate AI-related terminology in their name or business activities are classified as AI enterprises. And the data regarding the number of AI enterprises within the cities is sourced from statistics provided by the "Tianyan Check" website. Column (3) of Table 5.3 reveals a significant positive sign, which aligns with the above.

Table 5.3 Robust test results

	Alternative measure for ESG		*Alternative measure for AI*	*IV method*			
				IV for AI_install		*IV for AI_stock*	
	The Sino-Securities Index	*The Sino-Securities Index*	*ESG*	*Stage1: AI_install*	*Stage 2: ESG*	*Stage1: AI_stock*	*Stage2: ESG*
	(1)	*(2)*	*(3)*	*(4)*	*(5)*	*(6)*	*(7)*
AI_install	1.064** (0.495)				8.303*** (2.963)		
AI_stock		0.250** (0.109)					1.878*** (0.630)
The AI-enterprise density			0.462** (0.199)				
AI_install_IV				0.219*** (0.003)			
AI_stock_IV						1.186*** (0.013)	
Controls	Yes	Yes	Yes	Yes	Yes	Yes	Yes
Enterprise fixed-effect	Yes	Yes	Yes	Yes	Yes	Yes	Yes
Year fixed-effect	Yes	Yes	Yes	Yes	Yes	Yes	Yes
N	10,518	10,518	11,737	11,737	11,737	11,737	11,737
R^2	0.619	0.619	0.843		0.717		0.717
Kleibergen-Paap rk LM statistic				1265.19***		1195.74***	
Cragg-Donald Wald F statistic				32373.38***		35440.75***	
Kleibergen-Paap Wald rk F statistic				6127.74***		8467.29***	

*** $p < 0.01$; ** $p < 0.05$. Robust standard errors are in parentheses

Instrumental variable method The outcomes of regression analyses can be subject to bias if important variables are excluded from the model. To address this issue, we utilize the instrumental variable approach to reduce regression bias. In accordance with Zhang and Zeng (2024), this method utilizes data on industrial robots across various sectors in the United States as a proxy for data pertaining to Chinese industries, thereby enabling the creation of instrumental variables. Consequently, this chapter establishes AI_install_IV and AI_stock_IV as the instrumental variables for AI_install and AI_stock, respectively.

As a prominent entity in the field of AI, the United States has had a substantial impact on Chinese AI level. Nevertheless, the AI capabilities of the United States cannot directly affect Chinese enterprise ESG performance. Therefore, the use of these IVs is theoretically warranted. Additionally, this chapter includes an unidentifiable test and a weak instrumental variable test, the outcomes of which substantiate the statistical validity of the instrumental variables formulated in this chapter. As illustrated in Table 5.3, the findings indicate a beneficial influence of AI on enterprise ESG performance. Consequently, the findings derived from the instrumental variable method substantiate the reliability of the above baseline conclusions.

5.4.3 Heterogeneity Analysis

ESG item heterogeneity Enterprise ESG performance encompasses three distinct items: E, S, and G. The effects of AI on each item are presented in Table 5.4. The regression coefficients for E are statistically significant. This suggests that AI positively impacts enterprises' environmental responsibility. Concurrently, the regression results for S reveal that both AI_install and AI_stock exhibit significance, evidencing that AI can enhance their social responsibility performance.

In the case of G, the coefficients for AI_install and AI_stock do not meet the threshold for significance at the 10% level. Hence, AI has minimal influence on the governance levels of enterprises. One possible explanation for this finding is that G item is inherently more challenging to modify compared to E item and S item. Governance requires a systematic and holistic approach, which necessitates institutionalized modifications within enterprises. Additionally, governance metrics are frequently more challenging to quantify and substantiate compared to environmental and social indicators (Roger, 2024). As a result, AI may not sufficiently

Table 5.4 Results of heterogeneity analysis

	ESG item heterogeneity			*Region heterogeneity*		*Resource heterogeneity*		*Tech heterogeneity*		*Ownership heterogeneity*	
	E	*S*	*G*	*Western cities*	*Eastern and central cities*	*Resource-based cities*	*Non–resource-based cities*	*High-tech industries*	*Non–high-tech industries*	*SOEs*	*Non-SOEs*
	(1)	*(2)*	*(3)*	*(4)*	*(5)*	*(6)*	*(7)*	*(8)*	*(9)*	*(10)*	*(11)*
Part A: The independent variable is AI_install.											
AI_install	8.129*	8.828***	4.779	8.903	8.352***	43.975***	4.963*	20.154***	−0.568	11.172***	5.535
	(4.624)	(2.776)	(4.377)	(8.336)	(2.889)	(6.650)	(2.883)	(4.674)	(3.122)	(3.550)	(3.989)
R^2	0.655	0.754	0.809	0.836	0.844	0.839	0.844	0.848	0.848	0.851	0.842
Part B: The independent variable is AI_stock											
AI_stock	2.243**	2.030***	0.863	1.890	1.952***	9.647***	1.205*	4.423***	0.006	2.434***	1.476*
	(1.075)	(0.619)	(0.865)	(2.040)	(0.630)	(1.616)	(0.626)	(0.995)	(0.697)	(0.796)	(0.868)
R^2	0.655	0.754	0.809	0.836	0.844	0.840	0.844	0.848	0.848	0.851	0.842
Controls	Yes	Yes	Yes	Yes	Yes	Yes	Yes	Yes	Yes	Yes	Yes
Enterprise FE	Yes	Yes	Yes	Yes	Yes	Yes	Yes	Yes	Yes	Yes	Yes
Year FE	Yes	Yes	Yes	Yes	Yes	Yes	Yes	Yes	Yes	Yes	Yes
N	11,521	11,611	11,737	1763	9971	1380	10,353	4267	7449	6474	5224

*** $p < 0.01$; ** $p < 0.05$; * $p < 0.1$. Robust standard errors are in parentheses

incentivize enterprises to improve their governance practices through heightened oversight.

Region heterogeneity To test regional disparities, the samples are categorized into western cities, and eastern and central cities, and separate models are performed for them. Table 5.4 indicates that AI can improve the ESG performance of enterprises located in eastern and central cities; however, it does not hold true for western cities. Indeed, Chinese western cities exhibit a degree of underdevelopment, particularly concerning the understanding and implementation of ESG principles. This is especially evident in the E and S items, where there is a notable deficiency in effective external oversight mechanisms and a well-established institutional framework.

Resource heterogeneity Resource-based cities have established a development trajectory characterized by high input and low efficiency, primarily due to their abundant resource endowments, which has resulted in significant environmental degradation over an extended period (Zeng & Zhang, 2024). Hence, enterprises of resource-based cities possess the late mover advantage, enabling them to leverage AI to improve ESG performance more significantly. Hence, the sample is categorized according to the National Sustainable Development Plan for Resource-Dependent Cities,[6] followed by the implementation of regression analysis for each category independently. Their results are presented in Columns (6) and (7) of Table 5.4.

The findings are statistically significant for both resource-based and non–resource-based cities. This implies that AI enhances enterprise ESG performance in both types of cities. Notably, the coefficient for the former is substantially greater than that for the latter. Specifically, for the variable AI_install, the coefficient for resource-based cities is 43.975 (standard error: 6.650), whereas for non–resource-based cities, the coefficient is 4.963 (standard error: 2.883). Similarly, for the variable AI_stock, the former has a coefficient of 9.647 (standard error: 1.616), while the latter has a coefficient of 1.205 (standard error: 0.626). Hence, AI can improve enterprise ESG performance in resource-based cities to a greater extent.

[6]This document was released by the State Council of the People's Republic of China. Its URL is https://www.gov.cn/zwgk/2013-12/03/content_2540070.htm

Tech heterogeneity The high-tech industries exhibit a greater demand for technical components, suggesting that enhanced compatibility with AI technology may amplify AI's ESG performance enhancement effect. Hence, it investigates the tech heterogeneity. The regression coefficients for high-tech industries in Table 5.4 are significantly positive, whereas those for non–high-tech industries lack statistical significance. This indicates that AI can enhance enterprise ESG performance within high-tech industries, but does not yield a similar result in non–high-tech industries.

Ownership heterogeneity In contrast to non–state-owned enterprises (non-SOEs), SOEs in China bear a greater burden of environmental and social responsibilities. Consequently, SOEs may exhibit superior ESG performance, which highlights the possibility for AI to have a substantial impact on further advancing these enterprises' ESG performance.

The result for SOEs in Column (10) of Table 5.4 is significantly positive, suggesting AI can enhance SOEs' ESG performance. Conversely, Column (11) presents inconsistent results for non-SOEs, and the coefficient for AI_install does not meet the 10% significance threshold. Furthermore, the coefficients for SOEs are significantly higher than those for non-SOEs. Hence, compared to non-SOEs, AI exerts more pronounced effects on improving SOEs' ESG performance.

5.4.4 Underlying Mechanism

Analyst attention The advancement of AI enhances analysts' capacity to evaluate enterprises with respect to environmental, social, governance, and other dimensions, thereby augmenting their proficiency in acquiring and monitoring related information. This development compels enterprises to intensify their initiatives in these domains. Consequently, it can be posited that AI may impact enterprise ESG performance by influencing analyst attention. To investigate this mediating mechanism, this chapter constructs the following Eq. (5.3):

$$\text{Attention}_{jt} = \alpha_0 + \beta_1 \text{AI}_{jit} + \eta C_{jit} + u_j + v_t + \varepsilon_{jt} \tag{5.3}$$

where Attention represents analyst attention, which is quantified by the quantity of analysts monitoring the corresponding enterprises. The others align with Eq. (5.2).

Table 5.5 Results of underlying analysis

	Pressure: analyst attention		*Power: digital transformation*	
	(1)	*(2)*	*(3)*	*(4)*
AI_install	16.800***		3.834***	
	(5.404)		(0.476)	
AI_stock		4.438***		0.687***
		(1.183)		(0.100)
Controls	Yes	Yes	Yes	Yes
Enterprise FE	Yes	Yes	Yes	Yes
Year FE	Yes	Yes	Yes	Yes
N	11,737	11,737	11,737	11,737
R^2	0.678	0.679	0.791	0.791

*** $p < 0.01$. Robust standard errors are in parentheses

The coefficients of both AI_install and AI_stock in Table 5.5 are significantly positive. Hence, a greater presence of AI within a city correlates with increased scrutiny and interest from analysts toward local enterprises. In conclusion, AI can improve ESG performance by increasing analyst attention. Thus, Hypothesis 5.2 is supported. In fact, AI technological advancements boost analysts' productivity through real-time data analysis and advanced tools, enabling their precise ESG evaluations as well as automated reporting. And the assistance of AI to analysts enhances transparency to pressure enterprises toward ESG initiatives and incentivize enterprises to prioritize ESG performance.

Digital transformation As an advanced technological innovation, AI offers technical assistance, financial facilitation, and strategic direction to enterprises undergoing digital transformation. Concurrently, the enhancements in production efficiency and information processing capabilities resulting from this digital transformation have enabled enterprises to attain superior performance in ESG performance. To test the intermediary role of digital transformation, this chapter develops Eq. 5.4.

$$\text{Transformation}_{jt} = \alpha_0 + \beta_2 \text{AI}_{jit} + \eta C_{jit} + u_j + v_t + \varepsilon_{jt} \tag{5.4}$$

where Transformation represents digital transformation, which is measured by the frequency of keywords associated with digital transformation within enterprise annual reports, utilizing log form methodologies, drawing on Wu et al. (2021). The others align with Eq. (5.2).

These coefficients of AI_install and AI_stock in Table 5.5 are significantly positive. It means that AI has significantly facilitated the advancement of digital transformation within enterprises. Hence, AI can improve enterprises' ESG performance by accelerating their digital transformation, and Hypothesis 5.3 is supported. In fact, AI can help accelerate enterprises' digital transformation via technological isomorphism, enhancing data integration coupled with advancing digital technology adoption. Furthermore, AI-driven digital transformation helps elevate enterprise ESG performance by optimizing production chains, boosting resource efficiency, alongside reducing emissions, simultaneously fostering transparent management structures to align stakeholder interests.

5.4.5 Regulatory Analysis

Marketization level As the degree of marketization level increases, the behavior of enterprises becomes increasingly influenced by market competition. In environments characterized by high levels of marketization, enterprises tend to prioritize their public image, leading them to be more inclined to embrace their environmental and social responsibilities as a means of enhancing their competitiveness. This suggests that the marketization level may amplify AI's enhancement effect on ESG performance. To empirically evaluate this hypothesis, Eq. (5.5) has been developed for the purposes of this chapter.

$$\begin{aligned}\mathrm{ESG}_{jt} = \alpha_0 + \lambda_1 \mathrm{AI}_{jit} \times \mathrm{Marketization}_{jlt} + \beta_3 \mathrm{AI}_{jit} + \theta_1 \mathrm{Marketization}_{jlt} \\ + \eta C_{jit} + u_j + v_t + \varepsilon_{jt}\end{aligned} \tag{5.5}$$

where Marketization represents marketization level, which is gained from National Economic Research Institute (NERI).[7] l represents province and the others align with Eq. 5.2.

The coefficients of AI_install × Marketization and AI_stock × Marketization in Table 5.6 are significantly positive. Hence, as marketization increases, the influence of AI on enhancing enterprise ESG performance is expected to intensify. Hence, our findings support Hypothesis 5.4. In fact, marketization level acts as a catalyst for AI-driven ESG improvements by enabling freer resource flows, stronger property rights,

[7] NERI: http://www.neri.org.cn

and competitive markets. Specifically, advanced marketization attracts digital resources while reducing government intervention, forcing enterprises to adopt AI-enhanced ESG strategies. Complementing this, standardized ESG disclosure systems in mature markets create institutional pressures, thus embedding ESG priorities into corporate operations.

Business environment A conducive business environment facilitates enterprises in navigating a standardized legal framework and engaging in equitable market competition during their operational processes. This advantageous external context enhances the likelihood of enterprises to participate in ESG initiatives. Consequently, business environment may serve as a positive regulatory factor influencing the AI-ESG performance nexus. Hence, Eq. (5.6) is developed for the purpose of conducting relevant empirical tests.

$$\mathrm{ESG}_{jt} = \alpha_0 + \lambda_1 \mathrm{AI}_{jit} \times \mathrm{Business}_{jlt} + \beta_3 \mathrm{AI}_{jit} + \theta_1 \mathrm{Business}_{jlt} + \eta C_{jit} + u_j + v_t + \varepsilon_{jt} \tag{5.6}$$

where *Business* represents business environment, which is measured by the data from Wang et al. (2020). *l* represents province and the other symbols are consistent with those in Eq. 5.2.

The coefficients of both AI_install × Business and AI_stock × Business in Table 5.6 are significantly positive. These findings of evidence business environment serve as a supportive catalyst in AI' enhancement of ESG performance. Hence, our findings support Hypothesis 5.5. In fact, a robust business environment strengthens AI's role in advancing ESG performance by fostering efficient governance, innovation protection, and financial access. Meanwhile, these institutional synergies create a self-reinforcing cycle where AI technological adoption and business environment jointly amplify enterprises' ESG performance.

5.5 Discussions

5.5.1 Contributions

The contributions are as follows: Firstly, this chapter integrates AI-enterprise ESG performance nexus, examining the factors that drive enterprise ESG practices from an AI perspective. The findings carry significant practical meanings for advancing ESG practices within enterprises.

While prior studies have emphasized the importance of ESG practices for both enterprise and society (Chen & Xie, 2022; Qian, 2024), there is a notable scarcity of literature addressing its influencing factors, particularly regarding the direct influence of AI on enterprise ESG performance (Chen et al., 2024). Besides, we examined the AI-enterprise ESG performance at the city level, which has enhanced the robustness and depth of the findings.

Secondly, this chapter primarily highlights the noneconomic advantages of AI within the context of micro enterprises, thereby prompting a broader investigation into AI's impact on their behaviors. While the existing body of literature has largely concentrated on the economic benefits of AI—such as increased production capacity, optimized production processes, and enhanced market competitiveness (Babina et al., 2024)—there is a notable scarcity of research addressing the significant influence of AI on noneconomic behaviors within enterprises (Lee et al., 2025). This chapter aims to expand the academic discourse by exploring the effects of AI on noneconomic activities within enterprises.

Thirdly, this chapter elucidates the underlying mechanisms. Specifically, it posits that AI can enhance ESG performance by increasing analyst attention and facilitating digital transformation. Additionally, it suggests that marketization level and business environment may further amplify the positive effects of AI on enterprises' ESG performance. Consequently, this chapter contributes to a deeper understanding of the relationship between AI and enterprise ESG performance.

5.5.2 Policy Implications

Firstly, the government should prioritize the enhancement of AI and its role in improving enterprise ESG performance. This chapter provides empirical evidences for governmental initiatives focused on AI development. It is essential for the government to facilitate advancements in AI, offer institutional support for technological innovation, and foster a collaborative framework for AI development. Furthermore, the government should encourage the deeper integration of AI into enterprise ESG practices and reinforce the noneconomic responsibilities of enterprises in addressing social obligations, mitigating environmental degradation, and improving governance.

Secondly, the government should leverage AI capabilities to enhance analyst attention and facilitate digital transformation, thereby strengthening AI's contribution to enterprise ESG performance. It can be achieved

Table 5.6 Results of regulatory analysis

	Marketization level		*Business environment*	
	(1)	*(2)*	*(3)*	*(4)*
AI_install × Marketization	2.466*** (0.585)			
AI_stock × Marketization		0.458*** (0.132)		
AI_install × Business			16.668*** (6.339)	
AI_stock × Business				3.876** (1.598)
AI_install	0.879 (2.947)		−0.698 (3.901)	
AI_stock		0.458 (0.672)		−0.252 (0.963)
Marketization	−0.148* (0.090)	−0.167* (0.090)		
Business			0.856 (0.820)	0.874 (0.822)
Controls	Yes	Yes	Yes	Yes
Enterprise FE	Yes	Yes	Yes	Yes
Year FE	Yes	Yes	Yes	Yes
N	11,737	11,737	11,737	11,737
R^2	0.843	0.843	0.843	0.843

*** $p < 0.01$; ** $p < 0.05$; * $p < 0.1$. Robust standard errors are in parentheses

through supportive policies aimed at nurturing, developing, and expanding the pool of professional analysts, with a particular focus on enhancing their technical expertise in AI. Moreover, the government should allocate resources and establish policies that incentivize and assist enterprises in adopting digital technologies, particularly those exemplified by AI, to accelerate the transformation.

Thirdly, the government should prioritize marketization level and business environment to strengthen AI' effects on enterprise ESG performance. This involves minimizing administrative interventions in the market, facilitating the unrestricted flow of production factors, and fostering competitive dynamics among enterprises. Additionally, the government should improve business environment by streamlining administrative approval processes, reducing the tax and fee burdens on enterprises, reinforcing legal protections for property rights, and promoting transparency in governmental oversight.

5.6 Conclusions

Enterprise ESG practices benefit sustainable development; however, enterprises' ESG performance frequently falls short due to elevated costs and externalities. In recent years, AI has created new opportunities for enhancing enterprise ESG performance. Consequently, this chapter examines how AI affects enterprise ESG performance, utilizing a dataset comprising Chinese 1270 publicly listed enterprises across 187 cities from 2006 to 2019.

The findings indicate AI improves enterprise ESG performance across various robustness assessments. Besides, the enhancement effect exhibits heterogeneity. Specifically, AI demonstrates a significant capacity to enhance the Chinese enterprises' performance in E and S items, while it does not exert a notable effect on G item. Furthermore, this positive impact is particularly relevant to enterprises located in eastern and central cities, resource-based cities, high-tech industries, and SOEs. Additionally, the underlying mechanisms suggest that AI contributes to improved enterprise ESG performance by increasing analyst attention and facilitating digital transformation. Furthermore, marketization level and business environment can amplify the effectiveness of AI in enhancing enterprise ESG performance.

This chapter makes contributions to a deeper understanding of the AI–ESG nexus. However, certain limitations remain. Future research could be directed toward several key areas. First, how the implementation of AI technologies within enterprises impacts their ESG performance can be examined. As this chapter primarily considers the external urban AI level, future inquiries might focus on the internal application of AI within enterprises. Second, it is important to investigate how urban AI influences enterprise ESG investments. While this chapter emphasizes the outcomes of ESG initiatives, future research could consider the inputs involved in these processes.

References

Acemoglu, D., & Restrepo, P. (2020). Robots and jobs: Evidence from US labor markets. *Journal of Political Economy, 128*(6), 2188–2244. https://doi.org/10.1086/705716

An, Y., Deng, K. B., Ding, Z., & Gao, F. (2023). The role of marketization on the linkage between misallocation and productivity dispersion: Evidence from China. *Singapore Economic Review, 68*(3), 761–795. https://doi.org/10.1142/S0217590819500292

Babina, T., Fedyk, A., He, A. L., & Hodson, J. (2024). Artificial intelligence, firm growth, and product innovation. *Journal of Financial Economics, 151*, 103745. https://doi.org/10.1016/j.jfineco.2023.103745

Bai, F. P., Shang, M. T., & Huang, Y. J. (2024). Corporate culture and ESG performance: Empirical evidence from China. *Journal of Cleaner Production, 437*, 140732. https://doi.org/10.1016/j.jclepro.2024.140732

Cai, C., Tu, Y. Q., & Li, Z. (2023). Enterprise digital transformation and ESG performance. *Finance Research Letters, 58*, 104692. https://doi.org/10.1016/j.frl.2023.104692

Chen, Z. F., & Xie, G. X. (2022). ESG disclosure and financial performance: Moderating role of ESG investors. *International Review of Financial Analysis, 83*, 102291. https://doi.org/10.1016/j.irfa.2022.102291

Chen, R. X., & Zhang, T. X. (2024). Artificial intelligence applications implication for ESG performance: Can digital transformation of enterprises promote sustainable development? *Chinese Management Studies*. https://doi.org/10.1108/CMS-11-2023-0653

Chen, J., Wang, N., Lin, T. Z., Liu, B. L., & Hu, J. (2024). Shock or empowerment? Artificial intelligence technology and corporate ESG performance. *Economic Analysis and Policy, 83*, 1080–1096. https://doi.org/10.1016/j.eap.2024.08.004

Deng, Y., Jiang, W. Y., & Wang, Z. Y. (2023). Economic resilience assessment and policy interaction of coal resource-oriented cities for the low carbon economy based on AI. *Resources Policy, 82*, 103522. https://doi.org/10.1016/j.resourpol.2023.103522

Fan, Z. M., Chen, Y., & Mo, Y. F. (2024). Management myopia and corporate ESG performance. *International Review of Financial Analysis, 92*, 103071. https://doi.org/10.1016/j.irfa.2024.103071

Fang, M. Y., Nie, H. H., & Shen, X. Y. (2023). Can enterprise digitization improve ESG performance? *Economic Modelling, 118*, 106101. https://doi.org/10.1016/j.econmod.2022.106101

Fu, C., Luo, D. Y., Zhang, J. S., & Li, W. X. (2025). Tax incentives, marketization level, and corporate digital transformation. *International Review of Economics & Finance, 97*, 103777. https://doi.org/10.1016/j.iref.2024.103777

Gao, W., & Liu, Z. B. (2023). Green credit and corporate ESG performance: Evidence from China. *Finance Research Letters, 55*, 103940. https://doi.org/10.1016/j.frl.2023.103940

Gao, D., Zhou, X. T., & Wan, J. (2024). Unlocking sustainability potential: The impact of green finance reform on corporate ESG performance. *Corporate Social Responsibility and Environmental Management, 31*(5), 4211–4226. https://doi.org/10.1002/csr.2801

Guo, X., Liu, H. R., Saleem, S. F., & Budu, R. A. (2024a). Empowering sustainable human development and economic growth: Evidence from developing world. *Romanian Journal of Economic Forecasting, 27*(2), 32–49.

Guo, K., Bian, Y., Zhang, D. Y., & Ji, Q. (2024b). ESG performance and corporate external financing in China: The role of rating disagreement. *Research in International Business and Finance, 69*, 102236. https://doi.org/10.1016/j.ribaf.2024.102236

Holmstrom, J. (2022). From AI to digital transformation: The AI readiness framework. *Business Horizons, 65*(3), 329–339. https://doi.org/10.1016/j.bushor.2021.03.006

Huang, Y. J., Liu, S. C., Gan, J. W., Liu, B. L., & Wu, Y. X. (2024). How does the construction of new generation of national AI innovative development pilot zones drive enterprise ESG development? Empirical evidence from China. *Energy Economics, 140*, 108011. https://doi.org/10.1016/j.eneco.2024.108011

Huang, J., Han, F. F., & Li, Y. (2025). The impact of institutional investors' ESG concerns on corporate ESG disclosure: Evidence from site visits. *Finance Research Letters, 76*, 106957. https://doi.org/10.1016/j.frl.2025.106957

Huo, H., & Wang, Y. (2023). Research on the evolutionary game of multi-agent collaborative supervision of enterprise green behavior driven by big data. *Frontiers in environmental. Science, 11*, 1326322. https://doi.org/10.3389/fenvs.2023.1326322

Imjai, N., Yordudom, T., Yaacob, Z., Saad, N. H. M., & Aujirapongpan, S. (2025). Impact of AI literacy and adaptability on financial analyst skills among prospective Thai accountants: The role of critical thinking. *Technological Forecasting and Social Change, 210*, 123889. https://doi.org/10.1016/j.techfore.2024.123889

Johnson, K. N. (2019). Automating the risk of bias. *George Washington law review, 87*(5), 1214–1271.

Khurram, M. U., Chen, L. F., Abedin, M. Z., Adu, D. A., & Lucey, B. (2024). ESG disclosure and internal pay gap: Empirical evidence from China. *International Review of Economics & Finance, 92*, 228–244. https://doi.org/10.1016/j.iref.2024.02.023

Kuai, Y. C., Wang, P. W., & Huang, G. L. (2025). Corporate ESG performance and credit misallocation: Evidence from China. *Research in International Business and Finance, 73*, 102621. https://doi.org/10.1016/j.ribaf.2024.102621

Lee, C. C., Lu, M. T., Wang, C. W., & Cheng, C. Y. (2024). ESG engagement, country-level political risk and bank liquidity creation. *Pacific-Basin Finance Journal, 83*, 102260. https://doi.org/10.1016/j.pacfin.2024.102260

Lee, C. C., Hussain, J., & Abass, Q. (2025). An integrated analysis of AI-driven green financing, subsidies, and knowledge to enhance CO_2 reduction efficiency.

Economic Analysis and Policy, 85, 675–693. https://doi.org/10.1016/j.eap.2024.12.021

Li, C. M., Xu, Y., Zheng, H., Wang, Z. Y., Han, H. T., & Zeng, L. E. (2023). Artificial intelligence, resource reallocation, and corporate innovation efficiency: Evidence from China's listed companies. *Resource Policy, 81*, 103324. https://doi.org/10.1016/j.resourpol.2023.103324

Li, S. Y., Younas, M. W., Maqsood, U. S., & Zahid, R. M. A. (2024a). Impact of AI adoption on ESG performance: Evidence from Chinese firms. *Energy & Environment*. https://doi.org/10.1177/0958305X241269041

Li, G. L., Ren, K. J., Qiao, Y. H., & Wu, B. (2024b). From framing to priming: How does media coverage promote ESG preferences of institutional investors. *Emerging Markets Review, 63*, 101220. https://doi.org/10.1016/j.ememar.2024.101220

Lin, B. Q., & Zhu, Y. T. (2025). Does AI elevate corporate ESG performance? A supply chain perspective. *Business Strategy and the Environment, 34*(1), 586–597. https://doi.org/10.1002/bse.3999

Liu, Y., Zhang, F. K., & Zhang, H. (2024). CEO foreign experience and corporate environmental, social, and governance (ESG) performance. *Business Strategy and the Environment, 33*(4), 3331–3355. https://doi.org/10.1002/bse.3647

Luo, Y. G., Cui, H. J., Zhong, H. Y., & Wei, C. H. (2023). Business environment and enterprise digital transformation. *Finance Research Letters, 57*, 104250. https://doi.org/10.1016/j.frl.2023.104250

Nollet, J., Filis, G., & Mitrokostas, E. (2016). Corporate social responsibility and financial performance: A non-linear and disaggregated approach. *Economic Modelling, 52*, 400–407. https://doi.org/10.1016/j.econmod.2015.09.019

Qian, S. M. (2024). The effect of ESG on enterprise value under the dual carbon goals: From the perspectives of financing constraints and green innovation. *International Review of Economics & Finance, 93*, 318–331. https://doi.org/10.1016/j.iref.2024.03.010

Qin, M., Wan, Y., Dou, J. Y., & Su, C. W. (2024a). Artificial Intelligence: Intensifying or mitigating unemployment? *Technology in Society, 79*, 102755. https://doi.org/10.1016/j.techsoc.2024.102755

Qin, L. H., Su, H. D., Su, C. W., & Qin, M. (2024b). The essential role of big data: Could it effectively mitigate non-performing loans? *Romanian Journal of Economic Forecasting, 27*(3), 140–160.

Roger, T. (2024). Do financial analysts care about ESG. *Finance Research Letters, 63*, 105289. https://doi.org/10.1016/j.frl.2024.105289

Song, M. L., Pan, H. T., Shen, Z. Y., & Tamayo-Verleene, K. (2024). Assessing the influence of artificial intelligence on the energy efficiency for sustainable ecological products value. *Energy Economics, 131*, 107392. https://doi.org/10.1016/j.eneco.2024.107392

Su, C. W., Mirza, N., Umar, M., Chang, T. Y., & Albu, L. L. (2022). Resource extraction, greenhouse emissions, and banking performance. *Resources Policy, 79*, 103122. https://doi.org/10.1016/j.resourpol.2022.103122

Su, C. W., Shao, X. F., Jia, Z. J., Nepal, R., Umar, M., & Qin, M. (2023a). The rise of green energy metal: Could lithium threaten the status of oil? *Energy Economics, 121*, 106651. https://doi.org/10.1016/j.eneco.2023.106651

Su, C. W., Liu, F. Y., Stefea, P., & Umar, M. (2023b). Does technology innovation help to achieve carbon neutrality? *Economic Analysis and Policy, 78*, 1–14. https://doi.org/10.1016/j.eap.2023.01.010

Taherizadeh, A., & Beaudry, C. (2023). An emergent grounded theory of AI-driven digital transformation: Canadian SMEs' perspectives. *Industry and Innovation, 30*(9), 1244–1273. https://doi.org/10.1080/13662716.2023.2242285

Vezeteu, C. D., Dinca, M. S., & Stanciulescu, R. I. (2024). The influence of ESG on financial performance. Evidence from a combined cluster and panel regression analysis. *Romanian Journal of Economic Forecasting, 27*(4), 82–104.

Wan, G. C., Zhang, W. K., & Li, C. (2024). How does low-carbon city pilot policy catalyze companies toward ESG practices? Evidence from China. *Economic Analysis and Policy, 81*, 1593–1607. https://doi.org/10.1016/j.eap.2024.02.036

Wang, L. N., & Cao, L. (2024). Analyst attention and corporate green innovation. *Finance Research Letters, 69*, 104924. https://doi.org/10.1016/j.frl.2023.104924

Wang, X. L., Fan, G., & Hu, L. P. (2020). *Business environment index for China's provinces 2020 report.* Social Sciences Academic Press (China). In Chinese.

Wang, W. L., Sun, Q., & Zheng, M. B. (2021). Marketization level, fiscal input, and rural commercial bank performance. *Emerging Markets Finance and Trade, 57*(14), 4105–4120. https://doi.org/10.1080/1540496X.2020.1803825

Wang, J., Shu, T., Zhao, W. J., & Zhou, J. X. (2022). Research on Chinese consumers' attitudes analysis of big-data driven price discrimination based on machine learning. *Frontiers in Psychology, 12*, 803212. https://doi.org/10.3389/fpsyg.2021.803212

Wang, J. K., Tian, H. W., & Zheng, P. H. (2025). The impact of artificial intelligence on corporate environmental investment. *IEEE Transactions on Engineering Management, 72*, 96–114. https://doi.org/10.1109/TEM.2024.3491940

Wu, F., Hu, H. Z., Lin, H. Y., & Ren, X. Y. (2021). Enterprise digital transformation and capital market performance: Empirical evidence from stock liquidity. *Journal of Management World, 37*(7), 130–144+10. In Chinese.

Yang, S. F., & Wu, H. N. (2022). The global organizational behavior analysis for financial risk management utilizing artificial intelligence. *Journal of Global Information Management, 30*(7), 292455. https://doi.org/10.4018/JGIM.292455

You, X. H., Chen, C., Peng, K., & Li, Y. B. (2025). ESG disclosure quality and cost of debt. *Emerging Markets Review, 64*, 101219. https://doi.org/10.1016/j.ememar.2024.101219

Yu, G. H. (2024). Digital transformation, human capital upgrading, and enterprise ESG performance: Evidence from Chinese listed enterprises. *Oeconomia Copernicana, 15*(4), 1465–1508. https://doi.org/10.24136/oc.3058

Zeng, M., & Zhang, W. K. (2024). Green finance: The catalyst for artificial intelligence and energy efficiency in Chinese urban sustainable development. *Energy Economics, 139*, 107883. https://doi.org/10.1016/j.eneco.2024.107883

Zhang, W. K., & Zeng, M. (2024). Is artificial intelligence a curse or a blessing for enterprise energy intensity? Evidence from China. *Energy Economics, 134*, 107561. https://doi.org/10.1016/j.eneco.2024.107561

Zhang, W. K., & Zeng, M. (2025). When AI Powers Sustainability: Divergent Paths for Enterprise ESG Performance in China. *Romanian Journal of Economic Forecasting.*, *28*(3), 24–43.

Zhang, W. K., Fan, H. X., & Zhao, Q. W. (2023a). Seeing green: How does digital infrastructure affect carbon emission intensity? *Energy Economics, 127*, 107085. https://doi.org/10.1016/j.eneco.2023.107085

Zhang, F., Yang, B., & Zhu, L. (2023b). Digital technology usage, strategic flexibility, and business model innovation in traditional manufacturing firms: The moderating role of the institutional environment. *Technological Forecasting and Social Change, 194*, 122726. https://doi.org/10.1016/j.techfore.2023.122726

Zhang, W. K., Luo, Q., & Lin, X. (2024). Breaking boundaries: The influence of social insurance collection system reform on corporate innovation strategies. *Applied Economics*. https://doi.org/10.1080/00036846.2024.2405202

Zhang, W. K., Fan, H. X., & Zeng, M. (2025). Unleashing the power of artificial intelligence: A game changer for urban energy efficiency in China. *Sustainable Cities and Society, 126*, 106372. https://doi.org/10.1016/j.scs.2025.106372

Zhu, N. P., Zhang, S. Y., & Zhou, Y. Y. (2023). Impacts of environmental regulation on corporate ESG performance - evidence from China. *Romanian Journal of Economic Forecasting, 26*(3), 62–82.

CHAPTER 6

AI-Driven Reduction in Enterprise Energy Intensity

6.1 Introduction

Artificial intelligence (AI), which involves the replication of human behaviors using cutting-edge technologies including big data and deep learning (Acemoglu & Restrepo, 2020), is gaining significant traction globally (Zeng & Zhang, 2024; Zhang & Zeng, 2024; Zhang et al., 2025). A notable example is the proliferation of industrial robots, as illustrated in Fig. 6.1.[1] The global installations of industrial robots have experienced substantial growth, with an average annual increase of 13.27% from 2012 to 2022 ($(553/159)^{1/10}-1$). Notably, China achieved a remarkable milestone by installing 290,258 units of industrial robots in 2022. AI has become an unstoppable force, exerting a profound influence on all aspects of society. Recently, the impact of AI on energy consumption has emerged as a prominent topic of discussion. Some studies suggest a strong correlation between the proliferation of AI and a significant surge in global energy consumption, warning of a potential energy crisis without prompt

[1] International Federation of Robotics (IFR): https://ifr.org/

The core content of this chapter is derived from this article: Zhang, W. K., & Zeng, M. (2024). Is artificial intelligence a curse or a blessing for enterprise energy intensity? Evidence from China. *Energy Economics*, *134*, 107561. https://doi.org/10.1016/j.eneco.2024.107561

W. Zhang et al., *Green Transformation in the Digital Age*,
https://doi.org/10.1007/978-981-95-9121 3_6

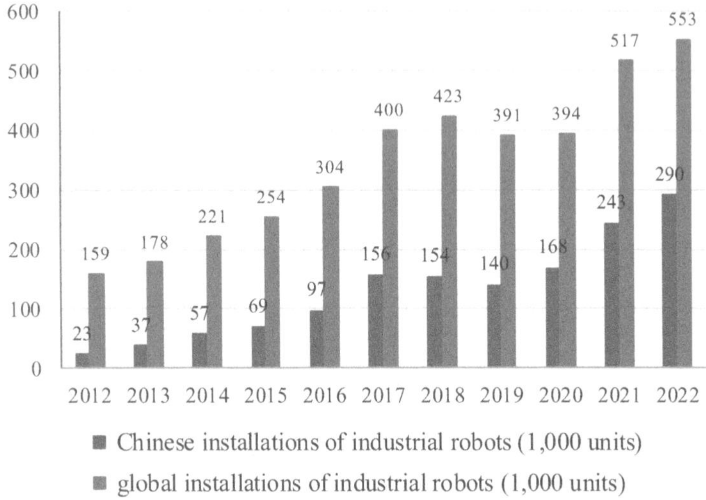

Fig. 6.1 Annual installations of industrial robots during 2012–2022

intervention (Dauvergne, 2022; Lee et al., 2025; Schwartz et al., 2020; Strubell et al., 2019). Conversely, other studies adopt a more optimistic stance, highlighting AI's potential to drive advancements in energy-saving technologies (Ding et al., 2024; Sarkar & Sarkar, 2020; Smajla et al., 2021; Zhang & Zeng, 2024).

In light of the global energy challenge, there is a pressing need to achieve energy sustainability (Gozgor & Paramati, 2022; Su et al. 2022a, b; Zhang & Zeng, 2024). The continuous increase in global energy consumption presents a gloomy outlook for energy sources. According to the Energy Institute,[2] global primary energy consumption in 2022 reached 604 exajoules, equivalent to 1.44 billion tonnes of oil equivalent. This represents an average annual growth rate of 1.4% between 2012 and 2022. Notably, China has emerged as the world's largest energy consumer since 2009, accounting for 26.39% of global primary energy consumption in 2022. Reducing energy intensity (EI), which refers to the amount of energy required per unit of output, is a crucial initiative in addressing energy challenges and achieving sustainability goals (Chishti et al., 2023; Zhang & Zeng, 2024). Therefore, it is crucial to explore the impact of AI on EI. However, the specific effects of AI on EI remain uncertain and require comprehensive consideration of the particular circumstances

[2] The Energy Institute: https://www.energyinst.org

(Zhang & Zeng, 2024). On the one hand, AI applications have the potential to enhance energy efficiency and minimize energy waste and loss, thus reducing EI. On the other hand, the development and utilization of AI necessitate significant computing resources and energy supply, which may increase energy consumption and subsequently raise EI.

The research related to the subject matter of this chapter can be categorized into two primary branches. The first branch focuses on the impacts of AI, with previous studies primarily examining its effects on technological progress (Jung & Lim, 2020), employment structure (Acemoglu & Restrepo, 2020), and economic growth (Berg et al., 2018). Makridis and Mishra (2022) concluded that emerging AI technologies drive job growth, and economic expansion, and consequently contribute to individual well-being. In addition to these macroeconomic impacts, existing research has also concentrated on analyzing the consequences of AI on micro-enterprises. For instance, Bork et al. (2023) emphasize that AI can enhance decision-making in enterprises and optimize the quality of decisions by developing and continuously updating language models to simulate and predict human behavior. Zdravkovic et al. (2022) provide evidence that AI has the potential to improve intelligent enterprise management, optimize enterprise information systems, and enhance efficiency in areas such as material scheduling, customer relationship management, and human resource management. Li (2021) finds that AI can facilitate capital flow management, enhance internal control mechanisms, and optimize corporate governance practices, thereby strengthening the resilience of the capital chain and fortifying organizations' ability to navigate external environmental disruptions. The second branch of research explores the determinants of EI. Existing studies have devoted significant attention to energy (Lee et al., 2022a; Lee et al., 2024; Zhang & Gu, 2023) and extensively investigated various factors that influence EI, including economic growth (Shahbaz et al., 2022), economic complexity (Payne et al., 2023), economic agglomeration (Wang et al., 2022), low-carbon city pilot policy (Wan et al., 2024), urbanization (Rafiq et al., 2016), trade openness (Chen et al., 2022), data factor marketization (Shen et al., 2025), and green innovation (Wurlod & Noailly, 2018). Additionally, studies suggest digital economy (Xu et al., 2022; Zhang et al., 2023) and digital inclusive finance (Shahbaz et al., 2023; Zhao et al., 2023) can contribute to the reduction of EI. Furthermore, Peng et al. (2022) demonstrate that moderate environmental regulation is conducive to helping enterprises reduce EI, whereas both excessively lenient and excessively stringent environmental regulations can promote their EI.

However, there is a lack of literature specifically addressing the relationship between AI and EI. The impact of AI on EI remains uncertain due to the trade-off between energy savings achieved by AI and the increased energy consumption it entails. Only three papers directly discuss the AI–EI relationship. Liu et al. (2021) provide evidence that AI contributes to the reduction of EI based on data from 16 Chinese industrial sectors spanning from 2006 to 2016. Similarly, Wang et al. (2022b) demonstrate the reduction effect of AI on EI using data from 38 countries and 17 manufacturing sectors. While these studies shed some light on the impact of AI on EI, they do not specifically analyze it from an enterprise perspective. Enterprises play a crucial role as economic agents, serving as important platforms for AI technology applications and significant contributors to energy consumption. Therefore, examining the micro-level impact of AI on EI from an enterprise perspective allows for a more comprehensive understanding. Additionally, investigating the AI–EI relationship at the industry level poses inherent challenges due to disparities in AI demand and energy intensity among industries. Analyzing this association at the enterprise level can partially mitigate these endogenous issues, thus enhancing the reliability of research findings. Huang et al. (2022), based on data collected from Chinese enterprises between 2010 and 2012, demonstrate that the adoption of robots in corporate operations effectively enhances energy performance and reduces EI. Their insights contribute to understanding the AI–EI nexus at the enterprise level, but concerns regarding the validity of their methodology exist. The authors have categorized enterprises into two groups: the treatment group, which consists of enterprises that have adopted robots, and the control group, which includes enterprises that have not adopted robots. They have utilized the difference-in-differences (DID) method to investigate whether the adoption of robots leads to a reduction in the enterprises' EI. However, the decision to adopt robots is endogenous, meaning that factors influencing this decision may also impact the enterprises' EI levels. Consequently, this treatment framework faces a significant endogeneity issue. Although the authors have made efforts to address this problem through Propensity Score Matching (PSM), the effectiveness of these measures in ensuring the reliability of their conclusions cannot be guaranteed.

This chapter utilizes data from Chinese manufacturing enterprises spanning the period from 2011 to 2019 to investigate the relationship between

AI and EI. AI is measured by the exposure of enterprises to industrial robots. This chapter finds that AI has a negative influence on EI. Specifically, the introduction of an additional unit of industrial robots per hundred workers leads to a reduction in EI by approximately 2.5%. These conclusions remain valid even after conducting various robustness tests, including replacing variables, using the doubly debiased Lasso (DDL) regression, and adopting the instrumental variable (IV) method. Moreover, this chapter examines the heterogeneous effect of AI on EI among various types of enterprises. It reveals that the reduction in EI due to AI is more pronounced in high energy-dependence enterprises, nonlabor-intensive enterprises, and state-owned enterprises (SOEs). Additionally, this chapter explores the underlying mechanisms through which AI affects EI. It demonstrates that AI contributes to the reduction in EI by advancing technological innovation and digital transformation. Finally, this chapter analyzes the role of business cycle, industrial concentration, and environmental regulation in the relationship between AI and EI. The results reveal that AI brings a greater reduction in EI during economic booms compared to recessions, while industrial concentration has a positive influence and environmental regulation has a negative influence on the reduction effect of AI on EI.

6.2 Theoretical Analysis

6.2.1 The Impact of AI on Enterprise EI

Enterprise EI refers to the quantification of energy consumed by an enterprise for each unit of output. This measure is calculated by dividing the total energy consumption of the enterprise by its revenue. The impact of AI on enterprise EI can be examined from two perspectives: its effect on enterprise revenue and energy consumption. Firstly, AI holds the potential to substantially increase enterprise revenue. More specifically, AI can enhance management efficiency and productivity. This can lead to a higher output per unit of labor (Calabrese et al., 2023). Furthermore, AI can generate new employment opportunities that offset the jobs displaced by automation (Acemoglu & Restrepo, 2019; Graetz & Michaels, 2018). In addition, AI can contribute to revenue growth by enhancing the value of products and expanding market share. Through the acceleration of technological progress, promotion of enterprise innovation, and product upgrades (Li et al., 2023), AI can augment the added value of products and improve the long-term competitiveness and market share of enterprises (Coccia, 2017).

Secondly, the impact of AI on enterprise energy consumption is multifaceted (Niu et al., 2025). On the one hand, AI can lead to increased energy consumption for enterprises. Specifically, AI itself is energy-intensive, meaning that when AI replaces human labor in production, enterprises will experience additional energy usage (Zhu et al. 2022). Moreover, AI can contribute to a rebound effect, further escalating energy consumption for enterprises (Gillingham et al., 2020; González, 2010). By automating repetitive tasks and enhancing efficiency, AI has significantly contributed to the expansion of production scale in various sectors such as manufacturing, logistics, and services. As a result, this has led to an augmented demand for energy to sustain the increased production. On the other hand, AI can also facilitate energy savings in enterprises through various methods. Cutting-edge AI techniques like deep learning and neural networks have displayed remarkable success in energy conservation (Lyu & Liu, 2021), particularly in reducing electricity consumption, which is a pivotal aspect of energy conservation (Ounifi et al., 2022). For instance, DeepMind, a subsidiary of Google, achieved a 40% reduction in energy consumption in 2016 by utilizing AI to predict fluctuations in electricity demand and optimize server and cooling system operations.[3] In fact, AI-enabled intelligent management systems can provide real-time monitoring of energy consumption across different departments and production processes, enabling immediate implementation of energy optimization schemes (Liao et al., 2023). Additionally, AI acts as a potent catalyst for the development of eco-friendly and energy-saving technologies (Liu et al., 2020), leading to its increasing adoption by enterprises to achieve energy conservation goals (Ying et al., 2023). Moreover, AI facilitates the implementation of a flat management structure within enterprises through the streamlining of management processes and the reduction of hierarchical levels. This enhancement significantly enhances management efficiency and ultimately contributes to energy conservation (Zdravković et al., 2022).

Existing studies suggest that the energy savings resulting from AI are projected to exceed its additional energy consumption. For example, several research studies, including Jin (2007), Vélez-Henao et al. (2020), and Adha et al. (2021), have confirmed that the rebound effect of energy conservation is lower than the actual conservation achieved. Moreover,

[3] Google DeepMind. https://www.deepmind.com/

Liu et al. (2021) have discovered that AI can effectively reduce EI while increasing economic output, as supported by data from 16 industries in China. Similar conclusions have been drawn by Chen et al. (2021) and Hossin et al. (2023). In summary, AI possesses the potential to enhance enterprise revenue while simultaneously decreasing energy consumption and lowering EI.

Therefore, this chapter proposes the following hypotheses:

Hypothesis 6.1: AI can significantly reduce enterprise EI.
Hypothesis 6.2: AI can reduce enterprise EI by increasing revenue and reducing energy consumption.

6.2.2 AI, Technological Innovation, and Enterprise EI

AI plays a crucial role in reducing enterprise EI by promoting technological innovation (Sun et al., 2021; Wurlod & Noailly, 2018; Zhang & Fu, 2022). By significantly expediting the process of knowledge acquisition and data processing, AI enables enterprises to swiftly innovate and introduce new technologies to their production systems. Technological innovation relies on the reorganization of existing knowledge and the creation of new knowledge (Chen, 2010), making the efficiency of knowledge acquisition and data processing crucial for innovation. AI technologies, such as advanced algorithms, empower enterprises to efficiently retrieve, identify, and analyze vast amounts of data, thereby facilitating the generation of fresh insights and the development of new knowledge (Liu et al., 2020; O'Leary, 2013). Consequently, AI accelerates the pace of technological innovation within enterprises, optimizing their energy consumption patterns and resulting in reduced energy consumption and enterprise EI (Lv et al., 2024).

Additionally, AI facilitates information exchange and fosters technical collaboration among enterprises, thus promoting their technological innovation. In modern society, technological innovation depends on the cooperation between multiple enterprises due to its inherent complexity. However, this cooperation is often impeded by physical and technical barriers. By overcoming these barriers, AI has opened up numerous opportunities for collaboration between enterprises. For instance, AI-enabled advancements such as tele-production, computer integration, and platform economics eliminate the constraints of distance and resources, thereby stimulating technological exchange and innovation (Holmstrom et al., 2016). In summary, AI can create a multitude of possibilities for inter-enterprise cooperation and drive technological progress, resulting in increased energy savings and reduced enterprise EI.

Finally, AI has become an essential catalyst for enhancing investments in technological innovation. It assumes a crucial role in cultivating an innovative culture within enterprises, both internally and externally. Internally, AI facilitates innovation investment by strengthening funding capabilities. By leveraging AI-powered automation, manufacturing processes can be streamlined, resulting in enhanced operational efficiency and productivity. Consequently, AI generates significant profits for enterprises, thereby incentivizing increased investment in research and development (R&D) (Lee et al., 2022b). Externally, AI promotes innovation investment by diversifying external financing channels for enterprises. By endowing enterprises with a competitive market edge and a robust corporate image, AI renders them more appealing to external investors. This is primarily because AI-driven innovations often lead to disruptive business models and cutting-edge products that can capture market share and create new value-generating opportunities (Shao et al., 2022). In conclusion, AI empowers enterprises to invest in technological innovation, thereby assisting them in fostering technological advancements and reducing their EI.

Therefore, this chapter proposes the following hypothesis:

Hypothesis 6.3: AI can reduce enterprise EI by promoting technological innovation.

6.2.3 *AI, Digital Transformation, and Enterprise EI*

"Digital transformation" is a significant avenue through which AI contributes to the reduction of EI for enterprises. This term refers to the process by which enterprises effectively utilize emerging information technologies, including AI, to optimize production processes and achieve a transformation in their business models (Matt et al., 2015; Vial, 2019). Numerous studies confirm the beneficial role of AI in facilitating digital transformation. For example, Lee et al. (2023) provide empirical evidence from the pension insurance sector, supporting the notion that AI promotes digital transformation. Similarly, Marina et al. (2023) demonstrate the positive impact of AI on digital transformation within the public health sector. Consistent findings are also evident in the field of marketing, as indicated by Srikrishna and Shivendra (2022).

On the one hand, AI has become an essential requirement for enterprises' digital transformation, particularly in the optimization of their production processes (Payne et al., 2021). The emergence of digital technologies has been the primary driver behind the escalation of enterprise production (Payne et al., 2021). The profound dependence of these digital technologies on AI makes it the most powerful tool in the enterprise transformation arsenal (Leone et al., 2021). As a result, AI has the potential to revolutionize enterprises by streamlining and optimizing their production, leading to improved energy utilization efficiency and reduced enterprise EI (Ben-Lahouel et al., 2021; Lin & Zhou, 2021).

On the other hand, AI plays a pivotal role in the evolution of modern enterprises by facilitating the transformation of business models. It empowers enterprises to digitize their business operations, office processes, and product offerings, thereby enabling the comprehensive digital transformation of their entire business model (Holmstrom, 2022). Therefore, AI undeniably acts as the most potent force in the business model transformation of enterprises, enabling them to establish energy-friendly production structures and business models that drive down their EI (Vassileva et al., 2012; Wang et al., 2022a). In conclusion, AI greatly benefits the realization of enterprises' digital transformation in production processes and business models, allowing for more efficient utilization of energy resources and a reduction in enterprise EI (Lin & Xie, 2023).

Therefore, this chapter proposes the following hypothesis:

Hypothesis 6.4: AI can reduce enterprise EI by promoting digital transformation.

Figure 6.2 summarizes the theoretical analysis framework of the impact of AI on enterprise EI.

6.3 Research Design

6.3.1 Data

This chapter aims to analyze the impact of AI on enterprise EI utilizing data from China's listed manufacturing enterprises from 2011 to 2019. The analysis encompasses three main data aspects. Firstly, this chapter

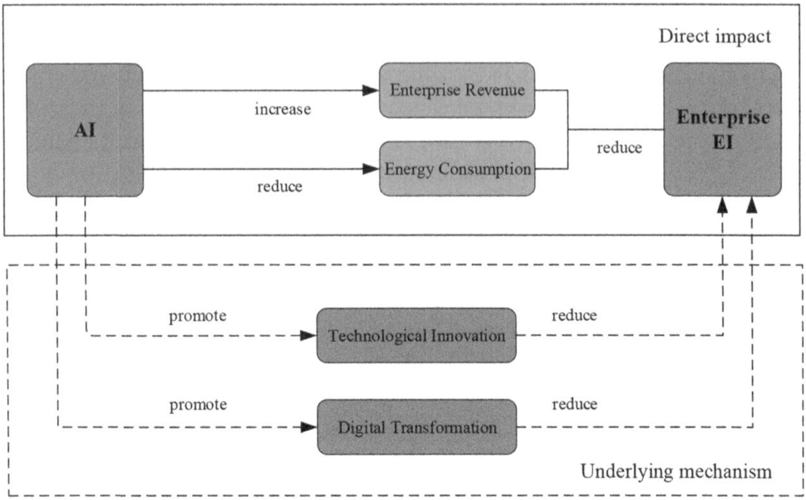

Fig. 6.2 Theoretical research framework

utilizes AI data, which is collected from the International Federation of Robotics (IFR).[4] The IFR has released data on the stock of robots by industry, country, and year from 1993 to 2019, which is widely used in AI-related research. However, it is important to note two considerations regarding the IFR data. Firstly, the IFR data divides manufacturing into 14 industries, while China's manufacturing encompasses 31 industries according to the Industrial Classification for National Economic Activities.[5] Therefore, this chapter aligns the IFR manufacturing industries with the Chinese manufacturing industries, following the approach employed by Wang and Dong (2020). Secondly, the IFR only provides industry-level data and not enterprise-level data. Consequently, this chapter disaggregates the IFR data from the industry level to the enterprise level, similar to the approach taken by Wang and Dong (2020). Given the scarcity of industrial robots in China before 2011, this chapter defines the sample period as 2011–2019. The second aspect of the analysis involves enterprise-level data. It is derived from the China Stock Market and Accounting

[4] IFR: https://ifr.org/

[5] Please visit China's National Bureau of Statistics website for more details, namely http://www.stats.gov.cn/

Research Database (CSMAR).[6] Lastly, city-level data is obtained from the Chinese Research Data Services Platform (CNRDS).[7]

6.3.2 Variable

The explained variable This chapter utilizes the framework proposed by Acemoglu and Restrepo (2020), Wang and Dong (2020), and Zhang and Zeng (2024) to measure AI by the exposure of enterprises to industrial robots (*AI*), namely the ratio of enterprises' industrial robot usage to their labor usage. *AI* is obtained by decomposing the exposure of industries to industrial robots ($\text{AI}_{\text{industry}}$), following the approach of Wang and Dong (2020). Specifically, *AI* is calculated by the product of $\text{AI}_{\text{industry}}$ and the decomposition weight ($\text{Weight}_{\text{AI}}$), which is defined in Eq. (6.1):

$$\begin{aligned} \text{AI} &= \text{AI}_{\text{industry}} \times \text{Weight}_{\text{AI}} \\ \text{AI}_{\text{industry}} &= \frac{R_{it}}{L_{i,t=2010}} \\ \text{Weight}_{\text{AI}} &= \frac{PWP_{ji,t=2011}}{\text{MedPWP}_{t=2011}} \end{aligned} \tag{6.1}$$

where $\text{AI}_{\text{industry}}$ refers to the exposure of industries to industrial robots; R_{it} represents the industrial robot usage of industry i in year t, while $L_{i,t=2010}$ represents the 2010 employment of industry i; *Weight* is the decomposition coefficient that measures the transition from industry-level exposure to enterprise-level exposure; $\text{PWP}_{ji,t=2011}$ denotes the proportion of production department workers in enterprise j in 2011, and $MedPWP_{t=2011}$ denotes the median value of all $\text{PWP}_{ji,t=2011}$.

The explanatory variable Enterprise EI is a metric that quantifies the energy consumption of an enterprise relative to its per unit of output (Chapple et al., 2013; Liu et al., 2021; Zhang & Zeng, 2024). Hence, this chapter calculates enterprise EI (*EI*) by taking the logarithm of enterprises' energy consumption divided by their main business revenue. The

[6] CSMAR: https://data.csmar.com/
[7] CNRDS: https://www.cnrds.com/

challenge lies in acquiring enterprises' energy consumption when measuring *EI*. Although the Industrial Enterprise Pollution Emissions Database releases Chinese enterprises' energy consumption data (Jiang et al., 2022), this data only covers the period up to 2012 and does not correspond to the research needs of this chapter. Due to the lack of data on enterprises' energy consumption, this chapter relies on estimations to assess the enterprises' energy consumption. In academia, two methods are commonly employed to estimate enterprises' energy consumption. The first method utilizes the energy expenditure costs of enterprises as a proxy for their energy consumption (Wen et al., 2021), while the second method involves decomposing the overall industry-level energy consumption to determine the energy consumption of individual enterprises (Shang et al., 2023). This decomposition method incorporates the Bartik instrumental variable approach, which can effectively estimate enterprises' energy consumption (Acemoglu & Restrepo, 2020). Hence, following Shang et al. (2023), enterprises' energy consumption is estimated by decomposing the industry-level energy consumption. To be specific, enterprises' energy consumption is estimated by the total energy consumption of the industry in which the enterprise is located multiplied by the proportion of the main business cost of the enterprise in the main business cost of the whole industry.

The control variables This chapter considers several enterprise-level and city-level confounding variables that could potentially impact the relationship between AI and enterprise EI. The enterprise-level variables include assets, the debt-to-asset ratio, the proportion of the leading three shareholders, the proportion of independent directors, and the annual salary of top-level executives. The city-level variables include per capita GDP; the ratio of fiscal revenue to GDP; the growth rate of GDP; and the proportion of the added value of the secondary industry in GDP.

Table 6.1 presents the descriptive statistics of the above variables. The mean value of *AI* is 0.418. This implies that Chinese enterprises maintain an average of 0.418 industrial robots per hundred individuals between 2011 and 2019. According to Wang and Dong (2020), from 2010 to 2015 Chinese enterprises maintain an average of 0.160 industrial robots per hundred workers. Hence, the measurement results of this chapter are in accordance with theirs, indicating that these results can be deemed

Table 6.1 Descriptive statistics

Variables	*Obs*	*Mean*	*S.D.*	*Min*	*Max*
Key variables					
Artificial intelligence (*AI*) (units/hundred workers)	15,043	0.418	0.961	0.000	8.480
Enterprise Energy intensity (*EI*) (in log, tons/million RMB)	15,043	2.333	1.026	0.004	5.241
Enterprise-level factors					
Assets (in log, RMB)	15,043	21.917	1.186	17.049	27.468
The debt to asset ratio	15,043	0.397	0.206	0.051	1.112
The proportion of the leading three shareholders	15,043	0.490	0.149	0.084	0.975
The proportion of independent directors	15,043	0.387	0.106	0	0.667
The annual salary of top-level executives (in log, RMB)	15,043	15.251	0.693	11.630	18.945
City-level factors					
Per capita GDP (in log, RMB)	15,043	11.541	0.789	8.851	13.185
The ratio of fiscal revenue to GDP	15043	0.105	0.043	0.024	0.239
The growth rate of GDP	15043	0.095	0.065	−0.374	0.614
The proportion of the added value of the secondary industry in GDP	15043	0.444	0.102	0.117	0.821

reliable. Meanwhile, these findings also indicate a substantial surge in the adoption rate of industrial robots within Chinese enterprises in recent years

6.3.3 *Method*

This chapter constructs Eq. (6.2) to test the impact of AI on enterprise EI, as given below:

$$EI_{jt} = \alpha_0 + \beta \mathrm{AI}_{jt} + \delta X_{jt} + \eta C_{jkt} + u_j + v_t + \varepsilon_{jt} \tag{6.2}$$

where *j*, *k*, and *t* represent enterprise, city, and year, respectively; EI is the dependent variable of enterprise EI; AI is the independent variable of AI, and its coefficient β reveals the effect of AI on enterprise EI; *X* and *C* represent the enterprise-level and city-level control variables, respectively, with coefficients δ and η; u_j and v_t incorporate the enterprise and year fixed-effects; α_0 denotes the intercepted item, while ε_{jt} is a normally distributed random error vector.

Table 6.2 Baseline regression results

	EI	*EI*	*EI*	*Energy-consumption*	*Revenue*
	(1)	*(2)*	*(3)*	*(4)*	*(5)*
AI	−0.022***	−0.025***	−0.025***	−0.444***	0.025*
	(0.003)	(0.003)	(0.003)	(0.020)	(0.014)
Enterprise-level controls		√	√	√	√
City-level controls			√	√	√
Enterprise FE	√	√	√	√	√
Year FE	√	√	√	√	√
Observations	15,043	15,043	15,043	15,043	15,043
R-squared	0.387	0.414	0.416	0.506	0.357

*** $p < 0.01$. Robust standard errors are in parentheses

6.4 Findings

6.4.1 Baseline Results

Table 6.2 presents the results of baseline regression analysis. Column (1) reports the results without any control variables, while Column (2) controls for enterprise-level variables. In Column (3), both enterprise-level and city-level variables are included. All columns introduce both enterprise and year fixed-effects.

The coefficients of all regressions are significantly negative at the 1% level, which indicates that AI can effectively reduce enterprise EI. These findings support Hypothesis 6.1. Regarding the specific coefficient values, there is no significant difference among the three regressions; thus, this chapter primarily focuses on the coefficient of Column 3, which includes control variables at both the enterprise and city levels. The coefficient for *AI* is −0.025, suggesting that an additional unit of industrial robots per hundred workers reduces enterprises' energy consumption per unit of main business income by approximately 2.5% ($e^{-0.025}-1\approx-0.025$). These findings are consistent with previous research conducted by Liu et al. (2021) and Wang et al. (2022b). Liu et al. (2021) find that a 1% increase in industrial robots in China's industrial sectors leads to an average decrease of 0.0244% in enterprise EI. Similarly, Wang et al. (2022b) demonstrate a significant average decrease of 0.0361% in enterprise EI following a 1% increase in industrial robots across global industrial sectors.

Despite the disparity in focus between this chapter, which examines the enterprise level, and their studies, which concentrate on the industry level, these consistent findings highlight the effectiveness of AI in mitigating enterprise EI.

Despite the significant energy consumption associated with AI, its potential for energy savings compensates for this aspect, making AI generally advantageous in reducing enterprise EI. Specifically, AI contributes to energy conservation through multiple means. Firstly, AI enables the monitoring and optimization of energy usage in enterprises. For instance, AI technologies like deep learning can be utilized to develop practical and dynamic models for energy monitoring and optimization (Feng et al., 2025; Liao et al., 2023). Secondly, AI-driven technological advancements have facilitated enterprises in achieving substantial increases in output while reducing energy input (Graetz & Michaels, 2018). Thirdly, AI enables intelligent management, effectively curbing energy consumption associated with internal communication in enterprises (Zdravković et al., 2022). As a result, this enhances energy efficiency and reduces enterprise EI.

The aforementioned evidence has demonstrated the impact of AI on reducing enterprise EI, which means that AI saves Chinese enterprises the amount of energy consumption for a unit of output. Three potential scenarios explain these outcomes. Firstly, AI may lead to an increase in both energy consumption and revenue for enterprises, and the increase in revenue outweighs the increase in energy consumption. Secondly, AI may result in a decrease in both energy consumption and revenue, with a more significant reduction observed in energy consumption. Thirdly, AI can simultaneously decrease energy consumption and increase revenue for enterprises. To validate these hypotheses, this chapter further examines the relationship between AI and energy consumption (measured by the log of enterprises' total energy consumption), as well as the relationship between AI and revenue (measured by the log of enterprises' main business revenue).

Their results are presented in Columns (4) and (5) of Table 6.2. The coefficient of AI on energy consumption is significantly negative, while that of AI on revenue is significantly positive. The findings of this chapter demonstrate that the implementation of AI technologies leads to a reduction in energy consumption among enterprises, while simultaneously contributing to an increase in their main business revenue. In summary, the adoption of AI enables enterprises to generate higher revenue while also promoting energy conservation, thereby mitigating their enterprise EI. As a result, these findings provide support for the third conjecture and validate Hypothesis 6.2.

6.4.2 *Robustness Tests*

Alternative measure for AI To ensure the credibility of the aforementioned findings, this chapter recalibrates the measurement of AI by modifying the decomposition weight in Eq. (6.1). The reconstruction of decomposition weight is based on the average values spanning from 2011 to 2019, and thus this chapter introduces a new measure of AI designated as AI_{robust}. The outcomes are presented in Column (1) of Table 6.3, revealing a highly significant negative coefficient at the 1% level. Consequently, these results align with the aforementioned analysis.

Table 6.3 Robustness test results

	Alternative measure for AI	*Alternative measure for enterprise EI*	*DDL regression*	*IV method*
	(1)	*(2)*	*(3)*	*(4)*
AI_{robust}	−0.024*** (0.003)			
AI		−0.025*** (0.003)	−0.014*** (0.003)	−0.008*** (0.002)
Enterprise-level controls	√	√	√	√
City-level controls	√	√	√	√
Enterprise FE	√	√	√	√
Year FE	√	√	√	√
Kleibergen-Paap rk LM statistic				415.410***
Cragg-Donald Wald F statistic				33000***
Kleibergen-Paap rk Wald F statistic				4018.043***
Observations	15,043	15,043	15,043	14,962
R-squared	0.416	0.435		0.414

*** $p < 0.01$ and ** $p < 0.05$. Robust standard errors are in parentheses

Alternative measure for enterprise EI This section reassesses enterprise EI by utilizing the logarithm of energy consumption divided by business revenue (denoted as EI_{robust}), instead of dividing it by main business revenue as previously done. The results in Column (2) of Table 6.3 reveal a significant negative coefficient for EI_{robust}, further confirming the role of AI in reducing enterprise EI.

DDL regression Despite efforts to control potential confounding factors affecting enterprise EI, this chapter may still overlook significant variables such as the energy supply structure of the city where the enterprise is located (Lau et al., 2023). The exclusion of these variables could introduce biases in the regression estimation results of enterprise EI by AI. Guo et al. (2022) propose the DDL method to mitigate the bias caused by missing variables, achieved through singular value compression of the regression element matrix. Referring to their research, this chapter adopts the DDL regression method to reexamine the impact of AI on enterprise EI. Column (3) of Table 6.3 reports the results of DDL regression. As shown, the coefficient *AI* is −0.014 and significant at the 1% level. Hence, according to the results of DDL regression, AI drives down enterprise EI.

IV method This chapter attempts to control confounding variables that may impact the relationship between AI and enterprise EI. However, it acknowledges that the issue of endogeneity, such as the inverse causality problem, may not be fully resolved. Therefore, this chapter employs the IV method to address this bias.

Consistent with Wang and Dong (2020), this chapter utilizes the exposure of industrial robots in US industries in 2010[8] as the IV for the exposure of industrial robots in Chinese industries.[9] Then the IV for *AI* is gained by multiplying the exposure of industrial robots in US industries in 2010 by the decomposition weight in Eq. (6.1). The choice of US industrial robot data in this chapter is supported by three key rationales (Wang & Dong, 2020). Firstly, the growth patterns of industrial robot usage in

[8] The US employment data by industry are acquired from the National Bureau of Economic Research (NBER) and the US Census Bureau's Center for Economic Studies (NBER-CES). NBER-CES: https://www.nber.org/esearch/da ta/nber-ces-manufacturing-industry-database

[9] There are differences between the US manufacturing industry classification and that of China. Based on the classification of China, this chapter reclassifies the U.S. manufacturing industry following Wang and Dong (2020).

both the United States and China demonstrate a significant degree of similarity. Secondly, the United States holds a prominent global position in industrial robot applications, serving as a reliable indicator of technological advancements in the industry. Lastly, in comparison to many developing nations, the manufacturing infrastructure in the United States is notably extensive. Hence, the US penetration of industrial robots is directly associated with the Chinese penetration of industrial robots, but it does not directly impact Chinese enterprise EI. Thus, the instrumental variable constructed in this chapter satisfies the necessary correlation and exclusivity requirements of IV methods. Furthermore, the under-identification and the weak instrumental variable tests (as shown in Table 6.3) are significant at the 1% level. In conclusion, both theoretical and empirical results support the suitability of the instrumental variable for AI. The results presented in Column (4) of Table 6.3 demonstrate that the coefficient of AI is significantly negative at the 1% level. This finding aligns with the previous results, confirming the robustness of our conclusions.

6.4.3 Heterogeneity Analysis

Energy dependence Enterprises exhibiting varying degrees of energy dependency demonstrate notable disparities in energy consumption and energy utilization efficiency. Industries that heavily rely on conventional energy sources, such as petrochemicals and steel, tend to demonstrate higher energy consumption and lower energy utilization efficiency. On the other hand, industries with lower energy dependence, such as services and internet-related sectors, have simpler production processes, which allows them to exercise better control over energy usage and enhance its utilization efficiency. Consequently, these industries display lower levels of energy consumption. Therefore, enterprises that are substantial energy consumers have a greater potential for improving energy consumption efficiency (Wu & Lin, 2022). For instance, Wang (2023) found a more pronounced reduction in energy consumption in high energy-dependence enterprises, driven by significant technological advancements resulting from environmental protection policies. As one of the most advanced technologies, AI may potentially contribute to a substantial reduction in enterprise EI for high energy-dependence enterprises. This chapter aims to investigate the heterogeneous impact of AI on enterprise EI based on the energy dependence of enterprises, which is determined by their respective industries. Following the classification provided by China's National

Table 6.4 Heterogeneity analysisresults

	Energy dependence		*Factor intensity*		*Enterprise ownership*	
	High	*Low*	*Labor*	*Nonlabor*	*SOEs*	*Non-SOEs*
	(1)	*(2)*	*(3)*	*(4)*	*(5)*	*(6)*
AI	−0.744***	−0.018***	0.332***	−0.032***	−0.031***	−0.023***
	(0.071)	(0.003)	(0.028)	(0.003)	(0.005)	(0.003)
Controls	√	√	√	√	√	√
Enterprise FE	√	√	√	√	√	√
Year FE	√	√	√	√	√	√
Observations	3112	11,931	3167	11,876	4,320	1,0723
R-squared	0.365	0.463	0.627	0.383	0.415	0.429
P-values for SUR	0.000		0.000		0.001	

*** $p < 0.01$. Robust standard errors are in parentheses

Bureau of Statistics (NBS),[10] this chapter categorizes the enterprises from five industries with high energy-consuming[11] to the high energy-dependence category, while the remaining enterprises are classified as low energy-dependence enterprises.

The results are reported in Columns (1) and (2) of Table 6.4. The coefficients of AI are all significant at the 1% level for both high and low energy-dependence enterprises. However, the coefficient for high energy-dependence enterprises (−0.744) is considerably larger than that for low energy-dependence enterprises (−0.018). To confirm the significance of this disparity, a Seemingly Unrelated Regressions (SUR) test is conducted, yielding a p-value of less than 0.01. These results indicate that the reduction effect in enterprise EI brought by AI is more pronounced in high energy-dependence enterprises compared to low energy-dependence enterprises. This can be attributed to two factors. Firstly, high energy-dependence enterprises have a higher level of energy consumption, which provides a wider scope for AI to reduce enterprise EI. Secondly, these

[10] NBS: http://www.stats.gov.cn/hd/cjwtjd/202302/t20230207_1902277.html

[11] The five industries are processing of petroleum, coking, and nuclear fuel (C25), manufacturing of raw chemical materials and chemical products (C26), manufacturing of nonmetallic mineral products (C30), ferrous metal smelting and rolling industry (C31), and nonferrous metal smelting and rolling industry (C32).

enterprises often face greater pressure from the government to save energy, which enhances their motivation to utilize AI for enterprise EI reduction.

Labor intensity Labor-intensive industries, such as the textile industry, are characterized by a substantial workforce primarily consisting of low-skilled labor. The adoption of AI, particularly industrial robots, is expected to displace a significant portion of this labor force, resulting in unemployment among low-skilled workers (Liu et al., 2020). This displacement of the labor force by AI undermines its potential role in mitigating enterprise EI. In contrast, nonlabor-intensive industries have a higher demand for resources, including capital, technology, and knowledge. These industries allocate more resources to technological advancements and strive for the widespread implementation of AI. Consequently, nonlabor-intensive industries may experience a greater reduction in enterprise EI. Therefore, the impact of AI on enterprise EI is likely to be more pronounced for nonlabor-intensive enterprises. This chapter aims to provide evidence for this heterogeneous impact by categorizing enterprises into labor-intensive and nonlabor-intensive industries. Following Pu et al. (2023), enterprises belonging to these 13 industries[12] are categorized as labor-intensive enterprises, while others are defined as nonlabor-intensive enterprises.

Their estimated results are in Columns (3)–(4) of Table 6.4. The coefficients of AI exhibit inconsistent signs for labor-intensive and nonlabor-intensive enterprises. Specifically, the coefficient of AI for labor-intensive enterprises is significantly positive at the 1% level, whereas it is significantly negative for nonlabor-intensive enterprises. This discrepancy is statistically significant, as demonstrated by the corresponding SUR test ($p < 0.01$). Consequently, AI only decreases the EI of nonlabor-intensive enterprises, while AI increases the EI for labor-intensive enterprises. These findings contradict the claims made by Liu et al. (2021), who argued that AI

[12] The 13 industries are agricultural and sideline product processing (C13), food manufacturing (C14), textile industry (C17), textile garment and apparel industry (C18), leather, fur, feather, and related products and footwear manufacturing industry (C19), timber processing and wood, bamboo, rattan, palm and grass products manufacturing industry (C20), furniture manufacturing industry (C21), paper and paper products manufacturing industry (C22), printing and reproduction of recording media manufacturing industry (C23), manufacturing of cultural, educational, industrial, scientific, physical, and recreational goods (C24), rubber and plastics products manufacturing industry (C29), nonmetallic mineral products manufacturing industry (C30), and metal products manufacturing Industry (C33).

reduces enterprise EI in both labor- and nonlabor-intensive industries, with a more significant impact observed in labor-intensive industries. The inconsistency in conclusions may be attributed to the different perspectives taken in this chapter, which analyzes the issue from an enterprise standpoint, and Liu et al. (2021), who approach it from an industry perspective. In reality, the introduction of AI technology may lead to an increase in enterprise EI for labor-intensive enterprises, as it replaces labor and consequently increases energy consumption. On the other hand, AI can have a more substantial effect in reducing enterprise EI for nonlabor-intensive enterprises, as these enterprises are more inclined to adopt AI technology.

Enterprise ownership There are clear disparities between state-owned enterprises (SOEs) and non-SOEs in China regarding their functional positioning, behavioral patterns, and policy support (Cheng & Ng, 2023). These discrepancies can result in different effects of AI on enterprise EI. For instance, SOEs have greater access to financial support compared to non-SOEs (Shao et al., 2022), which may enable them to effectively leverage AI and further reduce their EI. Therefore, the objective of this chapter is to examine the heterogeneous impacts of AI on enterprise EI based on enterprise ownership.

The findings are presented in Columns (5) and (6) of Table 6.4. The coefficients of AI demonstrate a statistically significant negative trend for both SOEs and non-SOEs. These results provide evidence that the implementation of AI has the potential to decrease enterprise EI in both types of enterprises. Specifically, the coefficient value of AI for SOEs (−0.031) is smaller than that of non-SOEs (−0.023), and the SUR test confirms a significant difference between the two groups ($p < 0.01$). Therefore, this chapter concludes that the impact of AI on enterprise EI is more significant in SOEs compared to non-SOEs. Several potential explanations can account for this phenomenon. Firstly, SOEs typically enjoy better conditions, including greater access to financing and policy support, which facilitate the realization of the reduction effect of AI on enterprise EI, unlike non-SOEs. Secondly, SOEs have a responsibility to support employment, resulting in a reduced substitution of labor by AI when compared to non-SOEs. This, in turn, leads to a greater reduction in enterprise EI in SOEs.

Table 6.5 Results of underlying mechanism

	Technological innovation	*Digital transformation*
	(1)	*(2)*
AI	0.044**	0.066***
	(0.022)	(0.022)
Controls	√	√
Enterprise FE	√	√
Year FE	√	√
Observations	15,043	15,043
R-squared	0.038	0.300

*** $p < 0.01$, and ** $p < 0.05$. Robust standard errors are in parentheses

6.4.4 Underlying Mechanism

Technological innovation Previous research consistently confirms that technological innovation plays a vital role in reducing enterprise EI. By improving production technology and restructuring production processes, technological innovation effectively enhances production efficiency and reduces energy consumption (Huang et al., 2017). Additionally, as an advanced technology, AI has the potential to empower enterprises in their innovation endeavors. Building upon this foundation, AI can contribute to the reduction of enterprise EI by promoting technological innovation. To investigate this proposed mechanism, this chapter primarily examines the impact of AI on technological innovation, considering the well-established and evident influence of technological innovation on reducing enterprise EI. Technological innovation is measured in this chapter using the logarithm of 1 plus the number of invention patent applications.

Column (1) of Table 6.5 displays the results about the underlying mechanism of technological innovation. The coefficient of *AI* is 0.044 and significantly positive at the 5% level. This implies that for every additional unit of industrial robots per hundred workers, there is an approximately 4.4% increase in enterprises' invention patent applications. This finding is consistent with the research conducted by Liu et al. (2020), which confirmed that AI influenced technological progress and its subsequent impact on enterprise EI at the industry level. These findings provide evidence that AI can stimulate enterprises' technological innovation, thereby supporting Hypothesis 6.3. The main reason for these findings is that AI can promote technological innovation through accelerating knowledge generation, facilitating technical cooperation, and increasing innovation investment.

Digital transformation Extensive research has confirmed the benefits of digital transformation in mitigating enterprise EI (Ben-Lahouel et al., 2021; Zhang et al., 2023). In fact, digital transformation not only enables enterprises to develop energy-efficient production structures and business models but also improves their operational efficiency. Moreover, AI plays a crucial role in driving the digital transformation of enterprises by ensuring the necessary prerequisites. Building on this foundation, AI has the potential to reduce enterprise EI by facilitating and promoting digital transformation. To examine this underlying mechanism empirically, this chapter aims to investigate the influence of AI on digital transformation, as previous research has already established the link between digital transformation and the reduction of enterprise EI.

The extent of digital transformation in enterprises is assessed by analyzing the frequency of keywords in the management discussion and analysis (MD&A) section of their annual reports. To be precise, this research compiles a list of 280 keywords related to digital transformation based on previous studies (Li et al. 2021; Wu et al., 2021; Zhao et al., 2021). Subsequently, the frequency of these keywords in the MD&A section is calculated, and a digital transformation index is constructed using the logarithm of their frequency.

The results are presented in Column (2) of Table 6.5, which indicates a statistically significant positive coefficient at the 1% level. These results demonstrate the potential of AI in expediting the digital transformation of enterprises, thereby supporting Hypothesis 6.4. These findings align with previous research that has identified the impact of AI on digital transformation across various domains, such as pension insurance (Lee et al., 2023), public health sector (Marina et al., 2023), and marketing (Srikrishna & Shivendra, 2022). Indeed, AI paves the way for enterprises' digital transformation by offering technical assistance (Payne et al., 2021; Leone et al., 2021). Given that digital technology serves as the core driving force behind the digital transformation of enterprises, AI, as a prominent digital technology, is indispensable in this process.

6.4.5 *Regulatory Analysis*

Moderating effect of business cycle Enterprises demonstrate unique behavioral patterns during various stages of the business cycle. For instance, during periods of economic booms, enterprises may strategically optimize their workforce, while during recessions, they may resort to downsizing. As a result, business cycle can potentially affect the decrease in enterprise EI caused by AI. Hence, this chapter aims to investigate whether the reduction in enterprise EI induced by AI varies across different phases of the business cycle. To accomplish this objective, we formulate the following Eq. (6.3):

$$EI_{jt} = \alpha_0 + \beta \mathrm{AI}_{jt} + \mu_1 \mathrm{Cycle}_{jkt} + \rho_1 \mathrm{AI}_{jt} \times \mathrm{Cycle}_{jkt} + \delta X_{jt} + \eta C_{jkt} + u_j + v_t + \varepsilon_{jt} \tag{6.3}$$

where Cycle represents business cycle. Cycle is assigned a value of 1 during booms and 0 during recessions, which is determined using the H-P filter method. The coefficient ρ_1 captures the impact of business cycle on the reduction effect of AI on enterprise EI. Other variables retain the same meaning as in Eq. (6.2).

The results are displayed in Column (1) of Table 6.6. The results reveal that the coefficient of *AI×Cycle* is significant and negative at the 1% level. These findings suggest that business cycle can amplify the reduction effect of AI on enterprise EI. Specifically, AI tends to generate larger reductions in enterprise EI during economic booms compared to recessions. One plausible explanation for this phenomenon is that, during periods of economic booms, enterprises typically operate at maximum production capacity and face high levels of demand. Consequently, they are more inclined to adopt AI technology to enhance production efficiency and reduce energy consumption (Begenau & Salomao, 2019; Zhu & Yuan, 2022). As a result, AI can assist businesses in mitigating enterprise EI by optimizing production processes, improving energy utilization efficiency, and forecasting energy demand, among other strategies (Zdravković et al., 2022). However, during economic recessions, enterprises encounter increased market pressure and a decline in demand. Consequently, they may decrease their investment in and utilization of AI technology (Huang & Rust, 2018). During such periods, enterprises may prioritize cost reduction and maintaining competitiveness instead of focusing on minimizing energy consumption.

Moderating effect of industrial concentration Industrial concentration plays a significant role in determining enterprise size, the level of external

Table 6.6 Regulatory analysis results

	Business cycle	*Industrial concentration*	*Environmental regulation*
	(1)	*(2)*	*(3)*
AI	−0.018***	−0.021***	−0.049***
	(0.003)	(0.003)	(0.007)
Cycle	0.010**		
	(0.004)		
AI×Cycle	−0.012***		
	(0.002)		
Concentration		0.009	
		(0.006)	
AI×Concentration		−0.008***	
		(0.002)	
Regulation			−0.012***
			(0.004)
AI×Regulation			0.006***
			(0.002)
Controls	√	√	√
Enterprise FE	√	√	√
Year FE	√	√	√
Observations	15,043	15,043	14,945
R-squared	0.417	0.417	0.417

*** $p < 0.01$, ** $p < 0.05$, and * $p < 0.1$. Robust standard errors are in parentheses

competitive pressure, and the extent of social responsibility, thus influencing the effectiveness of AI in reducing enterprise EI. The impact of AI on enterprise EI varies depending on the level of industrial concentration. The concentration of resources within an industry can be measured using the Herfindahl-Hirschman Index (HHI), which assesses resource concentration. Higher HHI values indicate concentration among a few enterprises, while lower values indicate dispersion across many enterprises (Hsu et al., 2014). This chapter aims to investigate the role of industrial concentration in the reduction effect of AI on enterprise EI, and to achieve this, the following equation, Eq. (6.4), is constructed:

$$EI_{jt} = \alpha_0 + \beta \mathrm{AI}_{jt} + \mu_2 \mathrm{Concentration}_{jkt} + \rho_2 \mathrm{AI}_{jt} \times \mathrm{Concentration}_{jkt} + \delta X_{jt} + \eta C_{jkt} + u_j + v_t + \varepsilon_{jt} \tag{6.4}$$

where *Concentration* represents industrial concentration. It equals the sum of the square of the ratio of the owners' equity of each enterprise in the industry to the total owners' equity of the industry. ρ_2 indicates its impact on the reduction effect of AI on enterprise EI.

The results pertaining to the moderating effect of industrial concentration are presented in Column (2) of Table 6.6. The findings reveal a significantly negative coefficient of *AI×Concentration* at the 1% level, indicating that the influence of AI on enterprise EI varies based on industry concentration. Specifically, in industries with high concentration, AI has a greater potential to reduce the EI of enterprises. This can be attributed to the fact that in highly concentrated industries, the application of AI technology enables enterprises to effectively manage and optimize energy usage, leading to a decrease in energy intensity. For example, in the manufacturing sector, AI can streamline production processes, minimize energy waste, and reduce inefficiencies. In contrast, in decentralized industries where energy consumption varies significantly among enterprises, the impact of AI applications on overall energy intensity is expected to be relatively limited. Additionally, industrial concentration also influences the extent to which enterprises adopt AI technology. In highly concentrated industries, the sharing of resources and data among enterprises is facilitated, simplifying the implementation of AI technologies. Conversely, decentralized industries face challenges in terms of data and resource sharing between enterprises, hindering the widespread adoption of AI.

Moderating effect of environmental regulation Environmental regulation refers to the government's imposition of restrictions on enterprises' energy consumption and pollution emissions through the formulation of policies and regulations (Shao et al., 2020; Zhang et al., 2022). Governments play a pivotal role in facilitating the energy transition and ensuring long-term energy sustainability (Sinha et al., 2023), and their environmental regulation effectively encourages enterprises to reduce energy consumption and improve energy efficiency (Du et al., 2020; Zhang et al., 2023). Consequently, environmental regulation may compel enterprises to address the energy consumption drawbacks of AI and enhance its potential to reduce enterprise EI. Therefore, this chapter aims to investigate the role of environmental regulation in the reduction of enterprise EI through AI. To achieve this objective, we construct the following equation, Eq. (6.5):

$$
\begin{aligned}
EI_{jt} = {} & \alpha_0 + \beta \mathrm{AI}_{jt} + \mu_3 \mathrm{Regulation}_{jkt} + \rho_3 \mathrm{AI}_{jt} \times \mathrm{Regulation}_{jkt} \\
& + \delta X_{jt} + \eta C_{jkt} + u_j + v_t + \varepsilon_{jt}
\end{aligned}
\tag{6.5}
$$

where *Regulation* represents environmental regulation is measured at the city level.[13] ρ_3 reflects the impact of environmental regulation, and the other signs are the same as those in Eq. (6.2).

Column (3) of Table 6.6 presents the results regarding the impact of environmental regulation. The results indicate a significant positive coefficient at the 1% level for the interaction term AI × Regulation. This suggests that environmental regulation can mitigate the reduction effect of AI on enterprise EI. Specifically, as environmental regulation becomes more stringent, the diminishing effect of AI on enterprise EI weakens. These findings contradict previous expectations, which have argued that environmental regulation drives enterprises to adopt various measures, including AI, to reduce their enterprise EI (Yin et al., 2023; Gao, 2023). However, our findings reveal that although the implementation of environmental regulation or AI can contribute to the reduction of enterprise EI (as shown in Column (3)), there exists a substitutive relationship between AI and environmental regulation in terms of reducing enterprise EI. In fact, the reduction in enterprise EI induced by environmental regulation can partially hinder enterprises' efforts to manage AI's energy consumption. On the one hand, achieving energy savings in AI can be costly due to the need for collaboration across different departments, processes, and stages (Haakman et al., 2021; Verdecchia et al., 2023). On the other hand, enterprises tend to comply with the minimum environmental requirements set by the government (Jiang et al., 2021).

6.5 Discussions

6.5.1 Contributions

The main contributions of this chapter are twofold. Firstly, this chapter significantly contributes to the field of AI and EI. Given the rapid expansion of AI and the growing energy constraints, it is crucial to investigate the impact of AI on EI. However, limited research has been conducted on

[13] Specifically, firstly, this chapter obtains the provincial-level environmental regulation with the number of environment-related proposals from the provincial Two Sessions, namely the Peoples Congress and the Political Consultative Conference, the two highest-ranking meetings. Secondly, this chapter obtains the city-level environmental regulation by multiplying the provincial-level one by the proportion of output value of city heavy industry to that of whole province, which is an example of Bartik decomposition, as Chen and Chen (2018) did. Thirdly, *Regulation* equals the log value of the city-level environmental regulation.

this topic, and the precise nature of their relationship remains uncertain. Therefore, this chapter fills a gap in the existing research by examining this issue from an enterprise perspective. Secondly, this chapter elucidates the mechanism through which AI affects enterprise EI, thereby enhancing our understanding of the relationship between AI and EI. Specifically, this chapter confirms that AI can reduce enterprise EI through technological innovation and digital transformation. Additionally, it provides a deeper understanding of the impact of AI on enterprise EI by considering factors such as business cycles, industrial concentration, and environmental regulation.

6.5.2 *Limitations*

While this chapter has yielded significant findings, there are still limitations that should be addressed in future research. Specifically, the current study focuses on the impact of AI on enterprise EI, but it does not analyze its effects on energy efficiency. EI, in this context, only takes into account a single input, energy consumption, and a single output, economic output, for enterprises. However, in reality, enterprises are faced with multiple inputs, such as energy, capital, and labor, as well as multiple outputs, including economic output and pollutant output. Thus, relying solely on EI may not provide a comprehensive understanding of enterprise energy performance. Additionally, due to constraints in available data, this chapter only examines the influence of AI on enterprise EI, rather than overall energy efficiency. To further investigate the role of AI in enhancing enterprise energy performance, future studies should explore its impact on energy efficiency once sufficient data becomes available.

6.5.3 *Policy Implications*

Based on the aforementioned findings, this chapter proposes several policy implications. Firstly, the government should formulate policies that promote the seamless integration of AI throughout the entire spectrum of enterprise production processes. This will enable enterprises to adopt environmentally friendly technologies, optimize production processes, and foster innovative business models, thereby effectively reducing enterprise EI. Secondly, the government should recognize that AI has varying effects on the EI of different types of enterprises. Therefore, customized policies should be formulated to amplify the role of AI in mitigating EI for each specific enterprise. Additionally, resources should be allocated to address the surge in EI faced by labor-intensive enterprises as a result of the emergence of AI. Thirdly, the government should strategically design policies according to the different stages of the business cycle. Specifically, during recessions, the government should actively promote employment to achieve an optimal balance between AI and labor, thereby preventing AI from becoming idle or undermining efficiency. Fourthly, the findings indicate that industrial concentration amplifies the reduction in enterprise EI achieved through AI. The government can gain insights from this to facilitate the systematic integration of enterprises and foster their expansion. Lastly, the government should prioritize the integration of mandatory methods with incentive-based approaches when promoting environmental protection policies. By combining these approaches, the government can effectively incentivize enterprises to enhance their motivation for reducing energy consumption and enterprise EI.

6.6 Conclusions

The rapid advancements in AI are revolutionizing various facets of enterprises at an unprecedented pace. Enterprises are leveraging the power of AI to optimize their operations, streamline work processes, and enhance customer experience, thereby enabling them to make informed decisions, improve productivity, and maintain a competitive edge. However, the impact of AI on enterprise EI remains uncertain, as it possesses both energy-saving potential and significant energy consumption itself. Given this context, this chapter set out to investigate the impact of AI on enterprise EI using Chinese manufacturing enterprises from 2011 to 2019. This chapter first identifies that the use of AI can effectively reduce enterprise EI. Specifically, an additional unit of industrial robots per hundred workers results in an approximately 2.5% decrease in energy consumption per unit of main business income. These conclusions remain robust after performing various tests, including alternative measures of AI and enterprise EI, using DDL regression and employing the IV method. Second, this chapter suggests that the reduction effect of AI on enterprise EI varies across enterprises with different characteristics. It is particularly pronounced in high energy-dependence enterprises, nonlabor-intensive enterprises, and SOEs. The underlying mechanism suggests AI reduces enterprise EI through advancing technological innovation and digital transformation within enterprises. Finally, this chapter reveals that business cycle, industrial concentration, and environmental regulation affect the effect of AI on enterprise EI. Specifically, AI brings a greater reduction of enterprise EI during economic booms and in industries with high concentration. There exists a substitutive relationship between AI and environmental regulation, and as environmental regulation becomes stricter, the decreasing effect of AI on enterprise EI weakens.

References

Acemoglu, D., & Restrepo, P. (2019). Automation and new tasks: How technology displaces and reinstates labor. *Journal of Economic Perspectives, 33*(2), 3–30. https://doi.org/10.1257/jep.33.2.3

Acemoglu, D., & Restrepo, P. (2020). Robots and jobs: Evidence from US labor markets. *Journal of Political Economy, 128*(6), 2188–2244. https://doi.org/10.1086/705716

Adha, R., Hong, C., Firmansyah, M., & Paranata, A. (2021). Rebound effect with energy efficiency determinants: A two-stage analysis of residential electricity consumption in Indonesia. *Sustainable Production and Consumption, 28*, 556–565. https://doi.org/10.1016/j.spc.2021.06.019

Begenau, J., & Salomao, J. (2019). Firm financing over the business cycle. *Review of Financial Studies, 32*(4), 1235–1274. https://doi.org/10.1093/rfs/hhy099

Ben-Lahouel, B., Taleb, L., Ben-Zaied, Y., & Managi, S. (2021). Does ICT change the relationship between total factor productivity and CO_2 emissions? Evidence based on a nonlinear model. *Energy Economics, 101*, 105406. https://doi.org/10.1016/j.eneco.2021.105406

Berg, A., Buffie, E. F., & Zanna, L. F. (2018). Should we fear the robot revolution? (the correct answer is yes). *Journal of Monetary Economics, 97*, 117–148. https://doi.org/10.1016/j.jmoneco.2018.05.014

Bork, D., Ali, S. J., & Dinev, G. M. (2023). AI-enhanced hybrid decision management. *Business & Information Systems Engineering, 65*(2), 179–199. https://doi.org/10.1007/s12599-023-00790-2

Calabrese, A., Costa, R., Tiburzi, L., & Brem, A. (2023). Merging two revolutions: A human-artificial intelligence method to study how sustainability and industry 4.0 are intertwined. *Technological Forecasting and Social Change, 188*, 122265. https://doi.org/10.1016/j.techfore.2022.122265

Chapple, L., Clarkson, P. M., & Gold, D. L. (2013). The cost of carbon: Capital market effects of the proposed Emission Trading Scheme (ETS). *Abacus, 49*(1), 1–33. https://doi.org/10.1111/abac.12006

Chen, C. J. (2010). The effects of knowledge attribute, alliance characteristics, and absorptive capacity on knowledge transfer performance. *R & D Management, 34*(3), 311–321. https://doi.org/10.1111/j.1467-9310.2004.00341.x

Chen, S. Y., & Chen, D. K. (2018). Air pollution, government regulations and high-quality economic development. *Economic Research Journal, 53*(2), 20–34. In Chinese.

Chen, C., Hu, Y. H., Karuppiah, M., & Kumar, P. M. (2021). Artificial intelligence on economic evaluation of energy efficiency and renewable energy technologies. *Sustainable Energy Technologies and Assessments, 47*, 101358. https://doi.org/10.1016/j.seta.2021.101358

Chen, S. S., Zhang, H. Y., & Wang, S. H. (2022). Trade openness, economic growth, and energy intensity in China. *Technological Forecasting and Social Change, 179*, 121608. https://doi.org/10.1016/j.techfore.2022.121608

Cheng, Q., & Ng, A. (2023). Achieving stability and prosperity: The Chinese way. *Humanities & Social Sciences Communications, 10*(1), 1–15. https://doi.org/10.1057/s41599-023-01656-4

Chishti, M. Z., Sinha, A., Zaman, U., & Shahzad, U. (2023). Exploring the dynamic connectedness among energy transition and its drivers: Understanding the moderating role of global geopolitical risk. *Energy Economics, 119*, 106570. https://doi.org/10.1016/j.eneco.2023.106570

Coccia, M. (2017). The source and nature of general purpose technologies for supporting next K-waves: Global leadership and the case study of the U.S. Navy's Mobile User Objective System. *Technological Forecasting and Social Change, 116*, 331–339. https://doi.org/10.1016/j.techfore.2016.05.019

Dauvergne, P. (2022). Is artificial intelligence greening global supply chains? Exposing the political economy of environmental costs. *Review of International Political Economy, 29*(3), 696–718. https://doi.org/10.1080/09692290.2020.1814381

Ding, T., Li, H., Liu, L., & Feng, K. (2024). An inquiry into the nexus between artificial intelligence and energy poverty in the light of global evidence. *Energy Economics, 136*, 107748. https://doi.org/10.1016/j.eneco.2024.107748

Du, W., Wang, F., & Li, M. (2020). Effects of environmental regulation on capacity utilization: Evidence from energy enterprises in China. *Ecological Indicators, 113*, 106217. https://doi.org/10.1016/j.ecolind.2020.106217

Feng, L. B., Qi, J. J., & Zheng, Y. H. (2025). How can AI reduce carbon emissions? Insights from a quasi-natural experiment using generalized random forest. *Energy Economics, 141*, 108040. https://doi.org/10.1016/j.eneco.2024.108040

Gao, Y. (2023). Unleashing the mechanism among environmental regulation, artificial intelligence, and global value chain leaps: A roadmap toward digital revolution and environmental sustainability. *Environmental Science*

and Pollution Research, 30(10), 28107–28117. https://doi.org/10.1007/s11356-022-23898-6
Gillingham, K., Rapson, D., & Wagner, G. (2020). The rebound effect and energy efficiency policy. *Review of Environmental Economics and Policy, 10*(1), 68–88. https://doi.org/10.1093/reep/rev017
González, J. F. (2010). Empirical evidence of direct rebound effect in Catalonia. *Energy Policy, 38*, 2309–2314. https://doi.org/10.1016/j.enpol.2009.12.018
Gozgor, G., & Paramati, S. R. (2022). Does energy diversification cause an economic slowdown? Evidence from a newly constructed energy diversification index. *Energy Economics, 109*, 105970. https://doi.org/10.1016/j.eneco.2022.105970
Graetz, Z., & Michaels, G. (2018). Robots at work. *Review of Economics and Statistics, 100*(5), 753–768. https://doi.org/10.1162/rest_a_00754
Guo, Z. J., Ćevid, D., & Bühlmann, P. (2022). Doubly Debiased LASSO: High-dimensional inference under hidden confounding. *The Annals of Statistics, 50*(3), 1320. https://doi.org/1214/21-AOS2152
Haakman, M., Cruz, L., Huijgens, H., & Van-Deursen, A. (2021). AI lifecycle models need to be revised. *Empirical Software Engineering, 26*(5), 1–29. https://doi.org/10.1007/s10664-021-09993-1
Holmstrom, J. (2022). From AI to digital transformation: The AI readiness framework. *Business Horizons, 65*(3), 329–339. https://doi.org/10.1016/j.bushor.2021.03.006
Holmstrom, J., Holweg, M., Khajavi, S. H., & Partanen, J. (2016). The direct digital manufacturing (r)evolution: Definition of a research agenda. *Operations Management Research, 9*, 1–10. https://doi.org/10.1007/s12063-016-0106-z
Hossin, M. A., Alemzero, D., Wang, R. P., Kamruzzaman, M. M., & Mhlanga, M. N. (2023). Examining artificial intelligence and energy efficiency in the MENA region: The dual approach of DEA and SFA. *Energy Reports, 9*, 4984–4994. https://doi.org/10.1016/j.egyr.2023.03.113
Hsu, S. Y., Lo, C. P., & Wu, S. J. (2014). The nexus of market concentration and privatization policy in mixed oligopoly. *Economic Modelling, 38*, 196–203. https://doi.org/10.1016/j.econmod.2013.12.025
Huang, M. H., & Rust, R. T. (2018). Artificial intelligence in service. *Journal of Service Research, 21*(2), 155–172. https://doi.org/10.1177/1094670517752459
Huang, J. B., Du, D., & Tao, Q. Z. (2017). An analysis of technological factors and energy intensity in China. *Energy Policy, 109*, 1–9. https://doi.org/10.1016/j.enpol.2017.06.048
Huang, G., He, L. Y., & Lin, X. (2022). Robot adoption and energy performance: Evidence from Chinese industrial firms. *Energy Economics, 107*, 105837. https://doi.org/10.1016/j.eneco.2022.105837

Jiang, Z. Y., Wang, Z. J., & Lan, X. (2021). How environmental regulations affect corporate innovation? The coupling mechanism of mandatory rules and voluntary management. *Technology in Society, 65*, 101575. https://doi.org/10.1016/j.techsoc.2021.101575

Jiang, Y. H., Zhang, Z. W., & Xie, G. J. (2022). Emission reduction effects of vertical environmental regulation: Capacity transfer or energy intensity reduction? Evidence from a quasi-natural experiment in China. *Journal of Environmental Management, 323*, 116180. https://doi.org/10.1016/j.jenvman.2022.116180

Jin, S. (2007). The effectiveness of energy efficiency improvement in a developing country: Rebound effect of residential electricity use in South Korea. *Energy Policy, 35*, 5622–5629. https://doi.org/10.1016/j.enpol.2007.05.028

Jung, J. H., & Lim, D. G. (2020). Industrial robots, employment growth, and labor cost: Asimultaneous equation analysis. *Technological Forecasting and Social Change, 159*, 120202. https://doi.org/10.1016/j.techfore.2020.120202

Lau, C. K., Gozgor, G., Mahalik, M. K., Patel, G., & Li, J. (2023). Introducing a new measure of energy transition: Green quality of energy mix and its impact on CO_2 emissions. *Energy Economics, 122*, 106702. https://doi.org/10.1016/j.eneco.2023.106702

Lee, C. C., Yuan, Z. H., & Wang, Q. R. (2022a). How does information and communication technology affect energy security? International evidence. *Energy Economics, 109*, 105969. https://doi.org/10.1016/j.eneco.2022.105969

Lee, C. C., Qin, S., & Li, Y. (2022b). Does industrial robot application promote green technology innovation in the manufacturing industry? *Technological Forecasting And Social Change, 183*, 121893. https://doi.org/10.1016/j.techfore.2022.121893

Lee, C. H., Wang, C., Fan, X. J., Li, F., & Chen, C. H. (2023). Artificial intelligence-enabled digital transformation in elderly healthcare field: Scoping review. *Advanced Engineering Informatics, 55*, 101874. https://doi.org/10.1016/j.aei.2023.101874

Lee, C. C., Yuan, Z. H., He, Z. W., & Xiao, F. (2024). Do geopolitical risks always harm energy security? Their non-linear effects and mechanism. *Energy Economics, 129*, 107245. https://doi.org/10.1016/j.eneco.2023.107245

Lee, C. C., Zou, J. Y., & Chen, P. F. (2025). The impact of artificial intelligence on the energy consumption of corporations: The role of human capital. *Energy Economics, 143*, 108231. https://doi.org/10.1016/j.eneco.2025.108231

Leone, D., Schiavone, F., Appio, F. P., & Chiao, B. (2021). How does artificial intelligence enable and enhance value co-creation in industrial markets? An exploratory case study in the healthcare ecosystem. *Journal of Business Research, 129*, 849–859. https://doi.org/10.1016/j.jbusres.2020.11.008

Li, R. (2021). Research on the impact of AI application on capital chain resilience. *Inzinerine Ekonomika-Engineering Economics, 34*(5), 536–553. https://doi.org/10.5755/j01.ee.34.5.33167

Li, G., Li, N., & Sethi, S. (2021). Does CSR reduce idiosyncratic risk? Roles of operational efficiency and AI innovation. *Production and Operations Management, 30*(7), 2027–2045. https://doi.org/10.1111/poms.13483

Li, C. M., Xu, Y., Zheng, H., Wang, Z. Y., Han, H. T., & Zeng, L. G. (2023). Artificial intelligence, resource reallocation, and corporate innovation efficiency: Evidence from China's listed companies. *Resources Policy, 81*, 103324. https://doi.org/10.1016/j.resourpol.2023.103324

Liao, H. T., Pan, C. L., & Zhang, Y. (2023). Smart digital platforms for carbon neutral management and services: Business models based on ITU standards for green digital transformation. *Frontiers in Ecology and Evolution, 11*, 1134381. https://doi.org/10.3389/fevo.2023.1134381

Lin, B. Q., & Xie, Y. J. (2023). Does digital transformation improve the operational efficiency of Chinese power enterprises? *Utilities Policy, 82*, 101542. https://doi.org/10.1016/j.jup.2023.101542

Lin, B., & Zhou, Y. (2021). Does the internet development affect energy and carbon emission performance? *Sustainable Production and Consumption, 28*, 1–10. https://doi.org/10.1016/j.spc.2021.03.016

Liu, J., Chang, H. H., Forrest, J. Y. L., & Yang, B. H. (2020). Influence of artificial intelligence on technological innovation: Evidence from the panel data of Chinas manufacturing sectors. *Technological Forecasting and Social Change, 158*, 120142. https://doi.org/10.1016/j.techfore.2020.120142

Liu, L., Yang, K., Fujii, H., & Liu, J. (2021). Artificial intelligence and energy intensity in China's industrial sector: Effect and transmission channel. *Economic Analysis and Policy, 70*, 276–293. https://doi.org/10.1016/j.eap.2021.03.002

Lv, T. G., Zhao, Q., Fu, S. F., Jin, G., Zhang, X. M., Hu, H., & Xu, G. L. (2024). Deciphering flows: Spatial correlation characteristics and factors influencing carbon emission intensity in the Yangtze River Delta. *Journal of Cleaner Production, 483*, 144290. https://doi.org/10.1016/j.jclepro.2024.144290

Lyu, W., & Liu, J. (2021). Artificial intelligence and emerging digital technologies in the energy sector. *Applied Energy, 303*, 117615. https://doi.org/10.1016/j.apenergy.2021.117615

Makridis, C. A., & Mishra, S. (2022). Artificial intelligence as a service, economic growth, and well-being. *Journal of Service Research, 25*(4), 505–520. https://doi.org/10.1177/10946705221120218

Marina, J., Abdullah, A., Antoine, H., & Salih, T. (2023). Digital transformation to mitigate emergency situations: Increasing opioid overdose survival rates through explainable artificial intelligence. *Industrial Management & Data Systems, 123*(1), 324–344. https://doi.org/10.1108/IMDS-04-2021-0248

Matt, C., Hess, T., & Benlian, A. (2015). Digital transformation strategies. *Business & Information Systems Engineering, 57*, 339–343. https://doi.org/10.1007/s12599-015-0401-5

Niu, X. T., Lin, C. G., He, S. S., & Yang, Y. C. (2025). Artificial intelligence and enterprise pollution emissions: From the perspective of energy transition. *Energy Economics, 144*, 108349. https://doi.org/10.1016/j.eneco.2025.108349

O'Leary, D. E. (2013). Artificial intelligence and big data. *IEEE Intelligent Systems, 28*(2), 96–99. https://doi.org/10.1109/MIS.2013.39

Ounifi, H. A., Gherbi, A., & Kara, N. (2022). Deep machine learning-based power usage effectiveness prediction for sustainable cloud infrastructures. *Sustainable Energy Technologies and Assessments, 52*, 101967. https://doi.org/10.1016/j.seta.2022.101967

Payne, E. H. M., Dahl, A. J., & Peltier, J. (2021). Digital servitization value co-creation framework for AI services: a research agenda for digital transformation in financial service ecosystems. *Journal of Research in Interactive Marketing, 15*(2), 200–222. https://doi.org.10.1108/JRIM-12-2020-0252

Payne, J. E., Truong, H. H. D., Chu, L. K., Dogan, B., & Ghosh, S. (2023). The effect of economic complexity and energy security on measures of energy efficiency: Evidence from panel quantile analysis. *Energy Policy, 177*, 113547. https://doi.org/10.1016/j.enpol.2023.113547

Peng, W., Lee, C. C., & Xiong, K. (2022). What shapes the impact of environmental regulation on energy intensity? New evidence from enterprise investment behavior in China. *Environmental Science and Pollution Research, 29*(35), 53364–53381. https://doi.org/10.1007/s11356-022-19655-4

Pu, X. H., Zeng, M., & Zhang, W. K. (2023). Corporate sustainable development driven by high-quality innovation: Does fiscal decentralization really matter? *Economic Analysis and Policy, 78*, 273–289. https://doi.org/10.1016/j.eap.2023.03.008

Rafiq, S., Salim, R., & Nielsen, I. (2016). Urbanization, openness, emissions, and energy intensity: A study of increasingly urbanized emerging economies. *Energy Economics, 56*, 20–28. https://doi.org/10.1016/j.eneco.2016.02.007

Sarkar, M., & Sarkar, B. (2020). How does an industry reduce waste and consumed energy within a multi-stage smart sustainable biofuel production system? *Journal of Cleaner Production, 262*, 121200. https://doi.org/10.1016/j.jclepro.2020.121200

Schwartz, R., Dodge, J., Smith, N. A., & Etzioni, O. (2020). Green AI. *Communications of the ACM, 63*(12), 54–63. https://doi.org/10.1145/3381831

Shahbaz, M., Song, M. L., Ahmad, S., & Vo, X. V. (2022). Does economic growth stimulate energy consumption? The role of human capital and R&D expenditures in China. *Energy Economics, 105*, 105662. https://doi.org/10.1016/j.eneco.2021.105662

Shahbaz, M., Siddiqui, A., Ahmad, S., & Jiao, Z. (2023). Financial development as a new determinant of energy diversification: The role of natural capital and structural changes in Australia. *Energy Economics, 126*, 106926. https://doi.org/10.1016/j.eneco.2023.106926

Shang, Y. P., Raza, S. A., Huo, Z., Shahzad, U., & Zhao, X. (2023). Does enterprise digital transformation contribute to the carbon emission reduction? Micro-level evidence from China. *International Review of Economics & Finance, 86*, 1–13. https://doi.org/10.1016/j.iref.2023.02.019

Shao, S., Hu, Z., Cao, J., & Guan, D. (2020). Environmental regulation and enterprise innovation: A review. *Business Strategy and the Environment, 29*(7), 1465–1478. https://doi.org/10.1002/bse.2446

Shao, J., Lou, Z. K., Wang, C., Mao, J. Y., & Ye, A. L. (2022). The impact of artificial intelligence (AI) finance on financing constraints of non-SOE firms in emerging markets. *International Journal of Emerging Markets, 17*(4), 930–944. https://doi.org/10.1108/IJOEM-02-2021-0299

Shen, N., Zhou, J. W., Zhang, G. P., Wu, L. J., & Zhang, L. (2025). How does data factor marketization influence urban carbon emission efficiency? A new method based on double machine learning. *Sustainable Cities and Society, 119*, 106106. https://doi.org/10.1016/j.scs.2024.106106

Sinha, A., Bekiros, S., Hussain, N., Nguyen, D. K., & Khan, S. A. (2023). How social imbalance and governance quality shape policy directives for energy transition in the OECD countries? *Energy Economics, 120*, 106642. https://doi.org/10.1016/j.eneco.2023.106642

Smajla, I., Sedlar, D. K., Vulin, D., & Juki, C. L. (2021). Influence of smart meters on the accuracy of methods for forecasting natural gas consumption. *Energy Reports, 7*, 8287–8297. https://doi.org/10.1016/j.egyr.2021.06.014

Srikrishna, C., & Shivendra, K. P. (2022). Artificial intelligence in marketing: A systematic literature review. *International Journal of Market Research, 64*(1), 38–68. https://doi.org/10.1177/14707853211101842

Strubell, E., Ganesh, A., & McCallum, A. (2019). Energy and policy considerations for deep learning in NLP. *57th AM ACL*, 3645–3650. https://doi.org/10.18653/v1/P19-1355.

Su, C. W., Yuan, X., Tao, R., & Shao, X. F. (2022a). Time and frequency domain connectedness analysis of the energy transformation under climate policy. *Technological Forecasting and Social Change, 184*, 121978. https://doi.org/10.1016/j.techfore.2022.121978

Su, C. W., Yuan, X., Umar, M., & Chang, T. Y. (2022b). Dynamic price linkage of energies in transformation: Evidence from quantile connectedness. *Resources Policy, 78*, 102886. https://doi.org/10.1016/j.resourpol.2022.102886

Sun, H., Edziah, B. K., Kporsu, A. K., Sarkodie, S. A., & Taghizadeh-Hesary, F. (2021). Energy efficiency: The role of technological innovation and knowledge spillover. *Technological Forecasting and Social Change, 167*, 120659. https://doi.org/10.1016/j.techfore.2021.120659

Vassileva, I., Wallin, F., & Dahlquist, E. (2012). Understanding energy consumption behavior for future demand response strategy development. *Energy, 46*, 94–100. https://doi.org/10.1016/j.energy.2012.02.069

Vélez-Henao, J. A., García-Mazo, C. M., Freire-González, J., & Vivanco, D. F. (2020). Environmental rebound effect of energy efficiency improvements in Colombian households. *Energy Policy, 145*, 111697. https://doi.org/10.1016/j.enpol.2020.111697

Verdecchia, R., Sallou, J., & Cruz, L. (2023). A systematic review of green AI. *Wiley Interdisciplinary Reviews. Data Mining and Knowledge Discovery, 13*(4), 1507. https://doi.org/10.1002/widm.1507

Vial, G. (2019). Understanding digital transformation: A review and a research agenda. *Journal of Strategic Information Systems, 28*(2), 118–144. https://doi.org/10.1016/j.jsis.2019.01.003

Wan, G. C., Zhang, W. K., & Li, C. (2024). How does low-carbon city polit policy catalyze companies toward ESG practices? Evidence from China. *Economic Analysis and Policy, 81*, 1593–1607. https://doi.org/10.1016/j.eap.2024.02.036

Wang, Y. F. (2023). Can the green credit policy reduce carbon emission intensity of "high-polluting and high-energy-consuming" enterprises? Insight from a quasi-natural experiment in China. *Global Finance Journal, 58*, 100885. https://doi.org/10.1016/j.gfj.2023.100885

Wang, Y. Q., & Dong, W. (2020). How the rise of robots has affected China's labor market: Evidence from China's listed manufacturing firms. *Economic Research Journal, 55*(10), 159–175. In Chinese.

Wang, Y. A., Yin, S. W., Fang, X. L., & Chen, W. (2022). Interaction of economic agglomeration, energy conservation and emission reduction: Evidence from three major urban agglomerations in China. *Energy, 241*, 122519. https://doi.org/10.1016/j.energy.2021.122519

Wang, Q. X., Hu, A., & Tian, Z. H. (2022a). Digital transformation and electricity consumption: Evidence from the Broadband China pilot policy. *Energy Economics, 115*, 106346. https://doi.org/10.1016/j.eneco.2022.106346

Wang, E. Z., Lee, C. C., & Li, Y. Y. (2022b). Assessing the impact of industrial robots on manufacturing energy intensity in 38 countries. *Energy Economics, 105*, 105748. https://doi.org/10.1016/j.eneco.2021.105748

Wen, H. W., Li, N. Y., & Lee, C. C. (2021). Energy intensity of manufacturing enterprises under competitive pressure from the informal sector: Evidence from developing and emerging countries. *Energy Economics, 104*, 105613. https://doi.org/10.1016/j.eneco.2021.105613

Wu, R. X., & Lin, B. Q. (2022). Environmental regulation and its influence on energy-environmental performance: Evidence on the Porter Hypothesis from China's iron and steel industry. *Resources, Conservation and Recycling, 176*, 105954. https://doi.org/10.1016/j.resconrec.2021.105954

Wu, F., Hu, H. Z., Lin, H. Y., & Ren, X. Y. (2021). Enterprise digital transformation and capital market performance: Empirical evidence from stock liquidity. *Journal of Management World, 37*(7), 15. In Chinese.

Wurlod, J. D., & Noailly, J. (2018). The impact of green innovation on energy intensity: An empirical analysis for 14 industrial sectors in OECD countries. *Energy Economics, 71*, 47–61. https://doi.org/10.1016/j.eneco.2017.12.012

Xu, Q., Zhong, M. R., & Li, X. (2022). How does digitalization affect energy? International evidence. *Energy Economics, 107*, 105879. https://doi.org/10.1016/j.eneco.2022.105879

Yin, K., Cai, F., & Huang, C. (2023). How does artificial intelligence development affect green technology innovation in China? Evidence from dynamic panel data analysis. *Environmental Science and Pollution Research, 30*(10), 28066–28090. https://doi.org/10.1007/s11356-022-24088-0

Ying, Y., Cui, X., & Jin, S. (2023). Artificial intelligence and green total factor productivity: The moderating effect of slack resources. *Systems, 11*, 356. https://doi.org/10.3390/systems11070356

Zdravković, M., Panetto, H., & Weichhart, G. (2022). AI-enabled enterprise information systems for manufacturing. *Enterprise Information Systems, 16*(4), 668–720. https://doi.org/10.1080/17517575.2021.1941275

Zeng, M., & Zhang, W. K. (2024). Green finance: The catalyst for artificial intelligence and energy efficiency in Chinese urban sustainable development. *Energy Economics, 139*, 107883. https://doi.org/10.1016/j.eneco.2024.107883

Zhang, R., & Fu, Y. (2022). Technological progress effects on energy efficiency from the perspective of technological innovation and technology introduction: An empirical study of Guangdong, China. *Energy Reports, 8*, 425–437. https://doi.org/10.1016/j.egyr.2021.11.282

Zhang, P. F., & Gu, H. Y. (2023). Potential policy coordination: Can energy intensity targets affect energy poverty? *Energy Economics, 126*, 106932. https://doi.org/10.1016/j.eneco.2023.106932

Zhang, W. K., & Zeng, M. (2024). Is artificial intelligence a curse or a blessing for enterprise energy intensity? Evidence from China. *Energy Economics, 134*, 107561. https://doi.org/10.1016/j.eneco.2024.107561

Zhang, W. K., Luo, Q., & Liu, S. Y. (2022). Is government regulation a push for corporate environmental performance? Evidence from China. *Economic Analysis and Policy, 74*, 105–121. https://doi.org/10.1016/j.eap.2022.01.018

Zhang, W. K., Fan, H. X., & Zhao, Q. W. (2023). Seeing green: How does digital infrastructure affect carbon emission intensity? *Energy Economics, 127*, 107085. https://doi.org/10.1016/j.eneco.2023.107085

Zhang, W. K., Fan, H. X., & Zeng, M. (2025). Unleashing the power of artificial intelligence: A game changer for urban energy efficiency in China. *Sustainable Cities and Society, 126*, 106372. https://doi.org/10.1016/j.scs.2025.106372

Zhao, C. Y., Wang, W. C., & Li, X. S. (2021). How does digital transformation affect the total factor productivity of enterprises? *Finance & Trade Economics, 42*(7), 114–129. In Chinese.

Zhao, H., Chen, S. W., & Zhang, W. K. (2023). Does digital inclusive finance affect urban carbon emission intensity: Evidence from 285 cities in China. *Cities, 142*, 104552. https://doi.org/10.1016/j.cities.2023.104552

Zhu, B., & Yuan, M. L. (2022). The business cycle and cost structure's adjustment speed. *Research in International Business and Finance, 62*, 101731. https://doi.org/10.1016/j.ribaf.2022.101731

Zhu, S., Ota, K., & Dong, M. (2022). Green AI for IIoT: Energy efficient intelligent edge computing for industrial internet of things. *IEEE Transactions on Green Communications and Network, 6*(1), 79–88. https://doi.org/10.1109/TGCN.2021.3100622

CHAPTER 7

Enterprise Innovation Leverage Via Green Credit

7.1 Introduction

In recent years, the Chinese government has implemented numerous policies aimed at promoting scientific and technological innovation. As a result, China has made significant scientific and technological advancements, and the impact of such progress on economic growth has increased considerably (Pu et al., 2023; Yang et al., 2024; Zhang et al., 2023). For instance, China's innovation output ranks 7th among 132 economies worldwide and 12th according to the 2021 Global Innovation Index.[1] However, China's innovation achievements are mainly reflected in the quantity dimension, and there are still deficiencies in quality innovation such as exploratory innovation (Lee et al., 2020; Zhang et al., 2023). The term "exploratory innovation" refers to an enterprise exploring potential opportunities, such as new products or services, and developing new markets by combining novel knowledge with existing knowledge elements

[1] The 2021 Global Innovation Index is published by the World Intellectual Property Organization (WIPO).

The core content of this chapter is derived from this article: Zhang, W. K., Luo, Q., Zhang, Y. F., & Yu, A. (2023). Does green credit policy matter for corporate exploratory innovation? Evidence from Chinese enterprises. *Economic Analysis and Policy*, *80*, 820–834. https://doi.org/10.1016/j.eap.2023.09.024

W. Zhang et al., *Green Transformation in the Digital Age*,
https://doi.org/10.1007/978-981-95-9121-3_7

(March, 1991; Zhang et al., 2023). It can help enterprises cope with environmental changes, especially new technological changes, and contribute to their industrial upgrading and long-term success (Wan et al., 2024; Wen et al., 2021b; Zhang et al., 2023). In response to these concerns, the Chinese government has issued a suite of policies, intending to improve the capacity for independent innovation (Zeng & Zhang, 2024; Zhang & Zeng, 2024; Zhang et al., 2023, 2025). Specifically, enterprises are being encouraged to perform "zero-to-one" exploratory innovation by absorbing novel knowledge (Gao et al., 2021; Cheng et al., 2025).

The green credit policy (GCP) is a crucial governmental regulation that plays an increasingly significant role in achieving the goals of green economy and carbon neutrality (Chiu & Lee, 2020; Liu et al., 2025a, b; Wang et al., 2024; Wen et al., 2022). Its implementation offers financial support for enterprises, reduces their innovation costs, and promotes corporate exploratory innovation (Wen et al., 2022). In 2012, the Chinese government issued the Green Credit Guideline (Guideline2012), which urges commercial banks to control environmental and social risks effectively in their credit business activities. The Guideline2012 requires banks to limit loans to enterprises with high environmental risks while increasing financial support for those that present low environmental risks (Wen et al., 2021a). This policy encourages environment-friendly enterprises to obtain more credit funds from commercial banks, thereby increasing their innovation activities (Lv et al., 2021b). Furthermore, less–environmentally friendly enterprises, particularly those in relatively low-tech, high-polluting traditional industries, are compelled to adopt eco-friendly practices. These actions enable them to capture new and emerging technological opportunities, leading to exploratory innovation for industrial upgrading. Therefore, investigating the influence of GCP on corporate exploratory innovation holds significant theoretical and practical value.

Numerous studies have documented the factors that influence corporate exploratory innovation, as well as the economic effect of GCP from various perspectives. On the one hand, some studies have identified numerous factors affecting corporate exploratory innovation, including top management team-related factors (Bai et al., 2021b; Su & Yang, 2018), knowledge and social networks (Guan & Liu, 2016; Zhang & Luo, 2020), financial slack and performance feedback (Hu et al., 2021b), and governmental policies (Zhang et al., 2020a). On the other hand, some studies have also investigated the economic effects of GCP (Du et al., 2023; Huang et al., 2022; Lee & Wang, 2022; Li & Chen, 2022; Shao et al., 2022; Song et al., 2021; Su et al., 2022; Zheng et al., 2023). Two pieces of literature directly related to this chapter, Hu et al.

(2021a) and Wang and Wang (2021), have addressed the impact of China's GCP on corporate green innovation. They found that GCP has a positive effect on corporate green innovation. Thus far, however, there is a dearth of literature on the impact of GCP on corporate exploratory innovation. Furthermore, the mechanisms of how GCP affects corporate exploratory innovation remain unknown.

This chapter aims to address the research gap by investigating the influence of GCP on corporate exploratory innovation in green credit-restricted industries, and the mechanisms through which this influence is exerted. The motivation for this chapter is twofold. On the one hand, as one of the largest and fastest-growing emerging countries in the world, China's development practice offers an exciting research topic due to its cultural and historical characteristics (Zhang et al., 2021). On the other hand, China faces some challenges and opportunities in green finance and high-quality innovation, including the need to balance green development and economic growth. Thus, exploring China's practice and experience in green finance and scientific and technological innovation can provide valuable reference and inspiration for other developing countries, and promote the process of global green finance and sustainable development.

To achieve this, this chapter uses the data of Chinese-listed enterprises from 2007 to 2017 to investigate the above issue in the following three main aspects. First, taking the Guideline2012 as a natural experiment, this chapter employs a propensity score matching and difference-in-differences (PSM–DID) method to examine the impact of GCP on corporate exploratory innovation. The findings reveal that GCP significantly promotes corporate exploratory innovation in green credit-restricted industries. To address the potential endogeneities, at least partially, this chapter performs some robustness tests, including a parallel trend test, alternative measures of variable, sample, and method, and a placebo test. Second, this chapter explores whether there is heterogeneity in the effect of GCP on corporate exploratory innovation based on enterprise characteristics such as ownership, industry, and region. The findings indicate that this positive impact mainly exists in state-owned enterprises (SOEs), high-capital-intensive industries and enterprises in China's central regions. Finally, this chapter sheds further light on the underlying mechanisms of GCP on corporate exploratory innovation from research and development (R&D)

investment and financing constraints. The findings suggest that the impact of GCP on corporate exploratory innovation becomes more significant as the increasing of financing constraints, and GCP can promote corporate exploratory innovation by increasing R&D investment.

7.2 Policy Background

In recent years, the Chinese government has issued numerous policies, laws, and regulations aimed at safeguarding the environment and controlling environmental pollution. Among these, GCP represents a critical component of the overall policy framework. Green credit encompasses a range of policies, systems, and practices designed to mitigate corporate pollution and enhance innovation and efficiency through credit intervention measures (Wen et al., 2021a). The GCP encourages credit institutions to actively implement green credit; it also provides preferential loans and subsidies, enabling enterprises to reduce their environmental risks.

In 2007, the Chinese government first introduced a GCP, which mandated financial institutions to adhere strictly to environmental protection policies in their credit activities. Subsequently, the Chinese government has enacted numerous laws and regulations pertaining to green credit. Notably, the Guideline2012 issued by the China Banking Regulatory Commission (CBRC) in 2012 is considered the most representative policy of the green credit system in China. The Guideline2012 delineates specific standards for commercial banks to undertake green credit and effectively manage environmental risks in their operations. The policy aims to facilitate green economic transformation by apportioning credit resources. Specifically, the Guideline2012 requires commercial banks to evaluate their customers' environmental and social risks and implement tailored dynamic credit-granting policies. Based on the outcome of the environmental and social risk assessment, commercial banks determine the appropriate credit-granting authority, including the loan amount, interest rate, and term. Conversely, credit will not be granted to enterprises that fail to comply with environmental and social performance standards. Since the implementation of the Guideline2012, green credit has witnessed rapid expansion in China.

7.3 Theoretical Analysis

7.3.1 GCP and Corporate Exploratory Innovation

Since March (1991) first proposed the concept of exploratory innovation, many researchers have extensively discussed this topic (e.g., Enkel & Gassmann, 2010; Zhang & Luo, 2020). Exploratory innovation is a process of searching or discovering new knowledge and introducing new methods or materials of knowledge dimensions into an organization (Guan & Liu, 2016; Karamanos, 2012; March, 1991). Its goal is to provide new designs, create new products, and develop new distribution channels for emerging customers or markets by broadening the existing knowledge base (Alexiev et al., 2010; Liu et al., 2025a, b). Undoubtedly, exploratory innovation can assist enterprises, particularly those in traditional industries with relatively low technological levels, to enter new technology fields and further promote technological progress and value growth. Whereas, exploratory innovation also encounters several barriers, such as high costs, high levels of uncertainty, and high risks (March, 1991), which can inhibit enterprises' motivation to carry out innovation activities. Moreover, abundant resource investments, such as external funding, play a prominent role in the corporate exploratory innovation process (Czarnitzki et al., 2011).

The GCP is a monetary policy regulation instrument that differs from command-based and market-based environmental regulation policies. Its main objective is to internalize the environmental pollution costs of enterprises operating in green credit-restricted industries by leveraging financing channels. The GCP provides an endogenous impetus for enterprise innovation by influencing the benefit–cost ratio of technological innovation. This policy imposes a cost on enterprises for their environmental pollution behaviors in terms of credit availability and loan financing costs, which subsequently raises their debt financing costs (Wang & Wang, 2021). Consequently, GCP encourages enterprises operating in green credit-restricted industries to prioritize green investments, reduce polluting investments, and promote green industrial transformation and upgrading through capital allocation.

The GCP might stimulate exploratory innovation activities among enterprises through the redistribution of capital elements (Wang & Wang, 2021). On the one hand, GCP has heightened the environmental access threshold of enterprise credit financing, thereby making credit funds available, convenient, and cost-effective for environmentally friendly enterprises. As a result, commercial banks can provide larger, longer-term, and lower-cost loans to enterprises with lower environmental and social risks (Wang & Wang, 2021). These enterprises are incentivized to continue engaging in exploratory innovation in order to keep enjoying incentives without penalties. These enterprises are incentivized to continue engaging in exploratory innovation activities as they can reap rewards without facing penalties.

On the other hand, GCP mandates that commercial banks restrict or refuse loans to enterprises with high environmental and social risks (Wen et al., 2021a). This policy imposes additional challenges and costs on these enterprises in obtaining loans, thereby limiting their credit resources. As a result, heavy-polluting enterprises may reduce their R&D investments and allocate limited funds to production and operational activities, hindering their ability to conduct exploratory innovation activities. In contrast, if these enterprises continue to maintain the original technology and production mode, they are bound to pay a price in financing availability and financing costs, which will damage their market competitiveness and impede the goal of profit maximization. Therefore, GCP may incentivize these enterprises to adopt more environmentally friendly green technologies through exploratory innovation to meet the requirements of green credit and avoid penalties (Wang & Wang, 2021). Additionally, GCP has a strong policy orientation, encouraging enterprises with high environmental and social risks to intensify innovation and transformation to improve overall productivity. This approach can offset the negative impact of green credit policies on economic profits and market competitiveness.

Therefore, this chapter proposes the following hypothesis:

Hypothesis 7.1: GCP significantly enhances corporate exploratory innovation in green credit-restricted industries.

7.3.2 Heterogeneity Analysis of GCP

Credit policies can have varying impacts on enterprises based on their ownership structure, industry, and regional characteristics (Aghion et al., 2012). This chapter aims to examine the heterogeneous impact of GCP on corporate exploratory innovation from three perspectives. First, the impact is heterogeneous between SOEs and non-SOEs. Specifically, there are significant differences between the two in terms of social responsibility, access to resources, and relations with the government (Kong et al., 2020; Kong et al., 2022). The credit-constrained industries are mainly dominated by SOEs, which have a national mission and are more aligned with the government's interests than non-SOEs. In response to the implementation of GCP, SOEs are more inclined to comply with national policies and achieve technological upgrades by implementing green credit policies. Moreover, commercial banks tend to favor property rights and scale (Yao et al., 2021). However, SOEs have natural political connections and diversified financing channels (Liu et al., 2021a), which are more likely to meet the green credit requirements. Therefore, SOEs may obtain large-scale, long-term, and low-cost loans from commercial banks to increase their investment in exploratory innovation activities. As a result, GCP may play a more significant role in encouraging SOEs to engage in exploratory innovation activities.

Second, according to traditional factor endowment theory, production factors are pivotal in the production activities of enterprises. Variations in factor input and production technology lead to significant differences in factor-intensive enterprises (Nguyen-Anh et al., 2022). Similarly, GCP has varying effects on exploratory innovation of high-capital-intensive and low-capital-intensive industries. High-capital-intensive industries are characterized by substantial investments, advanced technology and equipment, and minimal labor input. These enterprises require significant capital to carry out production and operation activities, heavily relying on bank credit funds. Conversely, low-capital-intensive industries rely heavily on labor and require fewer technology and equipment investments. Most enterprises in the green credit-restricted industries belong to high-capital-intensive industries. Therefore, GCP increases the difficulty and cost of obtaining credit for enterprises with high environmental and social risks. Consequently, GCP has a more noticeable impact on the exploratory innovation of high-capital-intensive enterprises with a tremendous demand for credit funds.

Third, there are various human capital and R&D capital endowments in China's eastern, central, and western regions, and their economic development shows apparent spatial aggregation and imbalance phenomenon (Bai et al., 2021a; Lv et al., 2021a). Specifically, the eastern region exhibits a higher degree of economic development and marketization compared to the central and western regions. Furthermore, innovation activities are concentrated mainly in the eastern region, which is indicative of a significantly higher innovation output than the other two regions.

Correspondingly, the effect of GCP on corporate exploratory innovation is likely to vary across regions. In other words, the impact of GCP on exploratory innovation may be heterogeneous due to regional differences in development levels. Enterprises in the western region face difficulties in meeting the requirements of GCP and obtaining loans from commercial banks for exploratory innovation activities due to shortcomings in financing channels, innovation resources, innovation capacity, and industrial upgrading. Conversely, enterprises in the other two regions, with strong profitability and high innovation levels, are more likely to meet the requirements of GCP and obtain funds for exploratory innovation activities.

Therefore, this chapter proposes the following hypotheses:

Hypothesis 7.2: GCP has a more significant positive effect on exploratory innovation of SOEs.

Hypothesis 7.3: GCP has a more significant positive effect on exploratory innovation of high-capital-intensive enterprises.

Hypothesis 7.4: GCP has a more significant positive effect on corporate exploratory innovation in the central region.

7.3.3 GCP, R&D Investment, and Corporate Exploratory Innovation

Exploratory innovation is often associated with high costs and risks and is typically constrained by limited resources (Jiang et al., 2018). R&D investment is a critical determinant of an enterprise's ability to achieve independent innovation and significantly impacts its innovation performance (Nemlioglu & Mallick, 2017; Nemlioglu & Mallick, 2021). The positive association between R&D investment and innovation has been well documented (Lee et al., 2021; Voutsinas et al., 2018), with R&D investment playing a vital role in exploratory innovation activities (Gao et al., 2021). Given the high risks and uncertainties associated with exploratory innovation, enterprises require substantial R&D investment to hire relevant professionals, procure raw materials and fixed assets, and cultivate and strengthen absorptive capacity (Wang et al., 2014).

The GCP is an environmental regulation policy that seeks to direct enterprises toward augmenting their R&D investment in production technology and enhancing their exploratory innovation ability (Zhuge et al., 2020). The policy offers long-term and low-cost loans to enterprises that exhibit good environmental performance. Conversely, it restricts loans to enterprises presenting high environmental and social risks. The policy's approach enables enterprises with commendable environmental performance to secure loans from commercial banks, thereby facilitating their exploratory innovation endeavors.

Therefore, this chapter proposes the following hypothesis:

Hypothesis 7.5: GCP promotes corporate exploratory innovation by increasing R&D investment.

7.3.4 GCP, Financing Constraints, and Corporate Exploratory Innovation

Financing constraints are a crucial factor that affects corporate innovation. Several studies have suggested that financing constraints can impose restrictions on corporate innovation activities (e.g., Aghion et al., 2012; Liu et al., 2021c). In particular, financially constrained enterprises tend to invest less in innovation or exploratory innovation due to the risk of failure (Tingbani et al., 2021). In China, commercial bank loans play a vital role in the financing channels of enterprises. However, the Guideline2012 has raised the threshold for enterprises to obtain credit financing from commercial banks. On the one hand, this policy has alleviated the financing constraints of enterprises with low environmental and social risks. These enterprises, as the leading practitioners of exploratory innovation, can access long-term and low-cost loans from commercial banks. These loans will alleviate the lack of investment in exploratory innovation typically caused by financing constraints, especially for those enterprises facing more severe financing constraints.

On the one hand, it is possible that GCP may worsen the financing difficulties faced by enterprises with high environmental and social risks. These enterprises, which are often heavily polluting, such as those in the steel and cement industries, tend to have high levels of debt and fixed asset investment. In situations where obtaining external financing from commercial banks is difficult, these enterprises may redirect funds that were originally intended for R&D and innovation toward more rigid expenditures, such as production and operation maintenance. This can lead to a crowding-out effect on financial support for R&D and innovation activities, which can negatively impact exploratory innovation. However, GCP may also incentivize these enterprises to adopt more environmentally friendly green technologies through exploratory innovation, in order to meet the requirements of green credit. This could potentially alleviate the financing constraints imposed by GCP. In light of these considerations, this chapter believes that GCP plays a significant role in promoting exploratory innovation among enterprises with high financing constraints.

Therefore, this chapter proposes the following hypothesis:

Hypothesis 7.6: GCP has a more significant promotion on corporate exploratory innovation as the degree of financing constraints increases.

7.4 Research Design

7.4.1 Data

This chapter employs a sample of Chinese-listed enterprises to examine the influence of GCP on corporate exploratory innovation using the PSM–DID method. The enterprise-level data is sourced from the China Stock Market and Accounting Research Database (CSMAR) (http://www.gtarsc.com), while the province-level data is collected from the China Statistical Yearbook. The innovation-related data is obtained from the Chinese research data services (CNRDS) (https://www.cnrds.com). In general, when utilizing the DID method to evaluate the policy effects, it is crucial to have a panel dataset of three to five years before and after the policy implementation. Given that the Guideline 2012 was implemented in 2012, this chapter considers the period from 2007 to 2017, including five years before and after the policy implementation. Besides, following Liu et al. (2021a, b) and Tian et al. (2020), this chapter processes the data based on the following criteria: (1) excluding enterprises with special treatment (ST)[2]; (2) excluding financial enterprises; (3) excluding enterprises from Growth Enterprise Market; (5) the linear interpolation method is applied to supplement some missing data; (6) continuous variables are winsorized at 1 and 99%.

[2]The Shanghai and Shenzhen Stock Exchanges of China give special treatment (ST) to listed enterprises with abnormal financial conditions.

7.4.2 Variable

The explained variable The dependent variable of this chapter is exploratory innovation (*ExpRatio*). Following the prior literature (e.g., Gao et al., 2021; Jia et al., 2019; Zhang et al., 2023), this chapter adopts the proportion of novel patent applications of an enterprise to the total number of invention patents to calculate exploratory innovation. As noted in earlier literature (e.g., Guan & Liu, 2016), four-digit International Patent Classification (IPC) subclass codes are commonly employed to identify an enterprise's entry into a new technology field and its exploratory innovations, as these codes reflect the heterogeneity of patents with respect to technical process, structure, and function. In line with Gao et al. (2021), a patent is defined as a novel if its IPC subclass code first appears in patents granted by SIPO within the previous four years. Given that invention patents undergo a more rigorous examination process compared to utility models and design patents (Jia et al., 2019), this chapter measures exploratory innovation by calculating the proportion of novel invention patents to the total number of invention patents in a given observation year. In robustness tests, exploratory innovation (ln*ExpNum*) is replaced by the natural logarithm of novel invention patent applications in a given observation year.

The explanatory variable The independent variable of this chapter is GCP (*Post*×*Treat*). *Post* is a dummy variable indicating the timing of the implementation of the Guideline2012. Since the Guideline2012 was implemented in 2012, *Post* equals 1 in 2012 and beyond, and 0 otherwise. *Treat* represents whether the sample enterprise belongs to green credit-restricted industries. Following Zhang et al. (2023), this chapter identifies the treatment group for green credit-restricted industries based on *the Key Evaluation Indicators of Green Credit Implementation* issued by the CBRC. Specifically, this chapter determines whether enterprises belong to the treatment group according to Class A and B industries in the *Key Evaluation Indicators*. The enterprises in the Class A and B industries are divided into the treatment group, and *Treat* is set to 1; otherwise, *Treat* is set to 0.

The control variables Consistent with prior literature conducted by Guan and Liu (2016) and Zhang et al. (2022), this chapter also incorporates a set of potentially confounding variables that may impact corporate exploratory innovation. These variables include both enterprise-level and province-level control variables. The enterprise-level control variables include the following: Enterprise size (ln*Asset*) is calculated by the natural log of total assets. Innovation expenditure (RDExp) is calculated by the ratio of R&D expenditure to main business revenue. Number of staff (ln*Nstaff*) is calculated by the natural log of the total number of staff. Executive salary (ln*DsmSalary*) is calculated by the natural log of the average executive salaries. Capital expenditure (ln*Capexp*) is calculated by the natural log of the capital expenditure. Return on equity (*ROE*) is calculated by the ratio of net profit to average stockholders' equity. Proportion of independent directors (*DrcRat*) is calculated by the ratio of independent directors to the total number of directors. Ownership concentration (*Top10*) is calculated by the proportion of shares held by the top 10% of shareholders. SOEs (*SOE*) equals 1 for SOEs, 0 otherwise. Chairman and CEO (*Dual*) equals 1 for the chairman's concurrent CEO position, 0 otherwise. The province-level control variables include the following: Economic development (ln*PGDP*) is calculated by the natural log of per capita GDP in each province. Fiscal revenue (*FisRev*) is calculated by the ratio of fiscal revenue to GDP in each province. Education expenditure (*EduExp*) is calculated by the ratio of education expenditure to GDP in each province. Tech expenditure (*TechExp*) is calculated by the ratio of technological expenditure to GDP in each province.

Table 7.1 presents the descriptive statistics for all variables. The results show that the mean value of *ExpRaito* is 0.30, with a corresponding standard deviation of 0.35. The mean value of ln*ExpNum* is 1.16, with a corresponding standard deviation of 1.19. These results indicate that exploratory innovation among Chinese enterprises is relatively low, with only 30% of invention patent applications being classified as exploratory innovation, and there are significant differences among the sample enterprises. The mean values of *Treat* and *Post* are 0.34 and 0.73, respectively, indicating that enterprises belonging to the treatment group account for 34% of the samples, and enterprises after implementing the Guideline account for 73% of the samples. The descriptive statistics for the other variables are generally within reasonable expectations. Moreover, a multicollinearity test is conducted to examine the potential issue of collinearity among the independent variables. The results reveal that the highest variance inflation factor (VIF) value among all the variables is 5.60, which is well below the commonly accepted threshold of 10. Therefore, it can be concluded that multicollinearity is not a concern among the variables in this chapter.

Table 7.1 Descriptive statistics

Variables	*Obs*	*Mean*	*S.D.*	*Median*	*Min*	*Max*	*VIF*
ExpRatio	11,834	0.30	0.35	0.16	0.00	1.00	–
lnExpNum	11,834	1.16	1.19	1.10	0.00	3.93	–
Treat	11,834	0.34	0.47	0.00	0.00	1.00	–
Post	11,834	0.73	0.45	1.00	0.00	1.00	–
Treat×post	11,834	0.24	0.43	0.00	0.00	1.00	1.09
lnAsset	11,834	22.17	1.27	21.99	19.13	26.03	3.66
RDExp	11,834	0.03	0.05	0.03	0.00	1.69	1.16
lnNstaff	11,834	7.95	1.16	7.88	3.83	11.06	2.63
lnDsmSalary	11,834	15.19	0.73	15.17	12.96	17.24	1.64
lnCapexp	11,834	0.11	0.15	0.07	0.00	1.15	1.10
Lev	11,834	0.44	0.20	0.44	0.06	0.90	1.61
ROE	11,834	0.06	0.13	0.07	−1.00	0.36	1.17
DrcRat	11,834	40.39	8.13	37.50	0.00	62.50	1.04
Top10	11,834	0.58	0.16	0.58	0.23	0.90	1.14
SOE	11,834	0.46	0.50	0.00	0.00	1.00	1.32
Dual	11,834	0.18	0.39	0.00	0.00	1.00	1.08
lnPGDP	11,111	6.06	2.55	5.87	1.11	12.90	3.26
FisRev	11,111	0.12	0.04	0.11	0.06	0.23	5.60
EduExp	11,111	0.04	0.01	0.04	0.02	0.17	1.77
TechExp	11,111	0.01	0.00	0.00	0.00	0.01	5.00

7.4.3 Models

This chapter utilizes the PSM–DID method to investigate the impact of GCP on corporate exploratory innovation. The DID model is constructed as Eq. (7.1):

$$Y_{it} = \alpha + \beta \text{Treat}_{it} \times \text{Post}_{it} + \lambda \text{Controls} + \delta_t + \theta_i + \varepsilon_{it} \tag{7.1}$$

where, Υ represents corporate exploratory innovation (*ExpRatio* and ln*ExpNum*); *i* and *t* represent enterprise and year, respectively; *Treat* × *Post* is the most concerned independent variable; *Controls* represents the control variables, including a set of enterprise-level and province-level control variables; δ and θ are enterprise and year fixed effects, and ε is the random error term.

7.5 Findings

7.5.1 Baseline Results

In this subsection, the PSM–DID method is utilized to investigate the impact of GCP on corporate exploratory innovation in China. Firstly, the PSM method is employed to match enterprises in the treatment and control groups, thus mitigating selection bias in the sample. Consistent with Zhang et al. (2020b), an equilibrium test is performed to ascertain that no systematic differences exist between the two groups. The results of the equilibrium test are presented in Table 7.2. The findings reveal that, prior to matching, significant differences in covariables between the two groups exist at a 5% level. However, after matching, none of the covariables exhibit statistical power, with absolute deviations below 10%. These results indicate that the two groups do not differ significantly after PSM.

Table 7.3 presents the baseline estimates using the PSM–DID method. Specifically, estimates with robust standard errors are reported in Columns (1) and (2), while estimates with clustering at the industry and city levels are reported in Columns (3) and (4).

One can find that the estimates of *Treat×Post* are all positive and statistically significant at the 1% or 5% levels. This finding provides evidence that GCP has significantly improved the exploratory innovation of enterprises

Table 7.2 Equilibrium test results

Variables	*Unmatched*	*Mean*		*Bias (%)*	*Reduced |bias|(%)*	*t-test*	
	Matched	*Treatment*	*Control*			*t-value*	*p-value*
lnAsset	U	22.239	21.964	21.20		4.29	0.000
	M	22.228	22.289	−6.90	67.60	−1.08	0.279
lnNstaff	U	7.952	7.652	22.80		4.55	0.000
	M	7.946	7.878	0.40	98.30	0.06	0.951
Lev	U	0.483	0.455	12.70		2.53	0.011
	M	0.480	0.487	−9.40	25.40	−1.55	0.121
ROE	U	0.058	0.071	−11.40		−2.37	0.018
	M	0.061	0.059	−7.00	38.40	−1.22	0.222

The table reports the equilibrium test of covariables using PSM. The covariables in 2012 (the year of the implantation of the Guideline 2012) are used to match the enterprises in the two groups. Covariables include lnAsset, lnNstaff, Lev, and ROE; U represents before PSM, and M represents after PSM. This chapter adopts the logit model and the 1:1 nearest matching method within propensity score caliper 0.01 to match samples

Table 7.3 Baseline regression results

	Dependent variable: exploratory innovation			
	(1)	*(2)*	*(3)*	*(4)*
Treat×Post	0.054***	0.046***	0.046***	0.046**
	(3.34)	(2.63)	(2.97)	(2.24)
Enterprise-level controls	Yes	Yes	Yes	Yes
Province-level controls	No	Yes	Yes	Yes
Enterprise FE	Yes	Yes	Yes	Yes
Year FE	Yes	Yes	Yes	Yes
Observations	11,834	11,111	11,111	11,111
Adj. R2	0.233	0.248	0.246	0.246

*** $p \le 0.01$, ** $p \le 0.05$, and * $p \le 0.10$; *t*-statistics are in parentheses

in green credit-restricted industries. Specifically, when the province-level control variables are included in Columns (2)–(4), the estimated effect remains statistically significant but is of a lower magnitude than the estimate for Column (1) without province-level control variables. Moreover, the estimated effects of GCP are still significant at the 1% or 5% levels, regardless of whether a robust standard error or clustering to industry and city is used in Columns (3) and (4). Thus, the results suggest that GCP significantly enhances exploratory innovation of enterprises in the green credit-restricted industries by approximately 4.60% (as demonstrated in Columns (2)–(4)), which supports Hypothesis 7.1.

These findings are similar to those of Hu et al. (2021a), Pang et al. (2025), and Wang and Wang (2021), and both demonstrate that GCP has a positive effect on corporate green innovation. However, Hu et al. (2021a) reported a significant increase of 61.7% in the green patent count for heavily polluting enterprises relative to nonheavily polluting enterprises following the promulgation of the Guideline2012. While Wang and Wang (2021) found that GCP results in a 21% rise in the total number of green patent applications in credit-restricted industries. This chapter is distinct from them in that it focuses on the perspective of corporate exploratory innovation. Compared to green innovation, corporate exploratory innovation is more challenging and time-consuming, resulting in a smaller growth rate.

The above findings can be attributed to the fact that GCP provides direction to enterprises for industrial transformation and upgrading via capital allocation (Wen et al., 2021a). When enterprises exhibit good environmental performance, they are better positioned to secure large-scale, low-cost, and long-term loans from commercial banks, which in turn enhances their risk tolerance and fosters exploratory innovation activities (Sharfman & Fernando, 2008).

As for the control variables, most of the estimates are consistent with the theoretical expectations, demonstrating that the findings are relatively reliable. Due to the potential endogeneity and the complexity of control variables, their estimates are not explained too much in this chapter.

7.5.2 *Robustness Tests*

Parallel trend test One vital assumption of DID method is that the two groups should satisfy parallel trends before implementing the Guideline2012. Namely, the changing trend of corporate exploratory innovation in the two groups should be similar. The regression coefficients and confidence intervals for the parallel trend test are presented in Fig. 7.1.

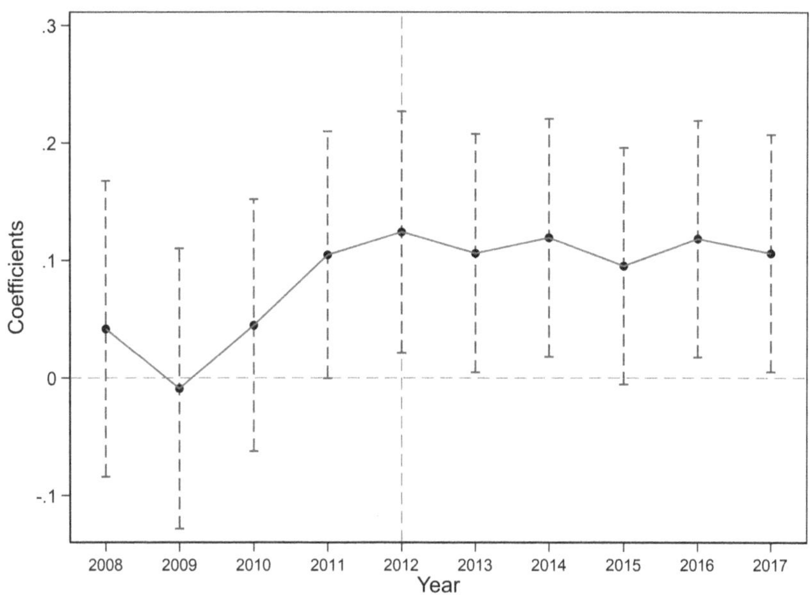

Fig. 7.1 Parallel trend test. (Notes: The circles represent the point estimates. Black dotted lines denote confidence intervals at the 95% level. The red dotted line represents the implementation year of GCP)

The figure shows that the estimates of the advance terms are insignificant at the 5% level before implementing the Guideline2012, while most of the coefficients of lag terms are significant after implementing Guideline 2012. This indicates that no significant difference exists in corporate exploratory innovation between the two groups before implementing the Guideline 2012. Therefore, the findings suggest that the parallel trend is generally satisfied.

Alternative measure for exploratory innovation In this section, this chapter employs the total number of novel patent applications (ln*ExpNum*) as a proxy for corporate exploratory innovation to examine the robustness of the key findings. The estimates are presented in Column (1) of Table 7.4. The results provide evidence that the estimate of *Treat*×*Post* is significant and positive at the 1% level, which further supports the notion that GCP has a positive impact on corporate exploratory innovation. This finding is in agreement with the main conclusions of this chapter.

Alternative definition for the treatment group Consistent with the study of Wang and Wang (2021), this chapter also constructs a new treatment group comprising enterprises classified under the Class A industries of *Key Evaluation Indicators*. $Treat_i$ equals 1 if an enterprise falls within the Class

Table 7.4 Robustness tests

	lnExpNum	*ExpRatio*		
	Alternative measure for exploratory innovation	*Alternative definition for treatment group*	*Eliminate enterprises in municipalities*	*Tobit method*
	(1)	*(2)*	*(3)*	*(4)*
Treat×Post	0.129***	0.039**	0.057***	0.067***
	(3.06)	(2.21)	(2.96)	(2.71)
Controls	Yes	Yes	Yes	Yes
Enterprise FE	Yes	Yes	Yes	Yes
Year FE	Yes	Yes	Yes	Yes
Observations	11,111	11,168	9268	11,151
Adj./Pseudo R2	0.572	0.248	0.248	0.367

*** $p \le 0.01$, ** $p \le 0.05$, and * $p \le 0.10$; *t*-statistics are in parentheses

A industries and 0 otherwise. The regression analysis is repeated using Eq. (7.1), and the estimates are presented in Column (2) of Table 7.4. The results demonstrate that GCP can significantly promote corporate exploratory innovation in green credit-restricted industries. This finding further reinforces the robustness and validity of this chapter's conclusions.

Eliminate the samples in municipalities In China, municipalities hold a higher administrative position than prefecture-level cities. Municipalities possess many advantages in terms of the economic scale, resource endowment, and innovation capacity, which may influence the impact of GCP on corporate exploratory innovation. To ensure the robustness of the above conclusions, this chapter excludes enterprises located in municipalities, such as Beijing, Shanghai, Tianjin, and Chongqing. The estimates presented in Column (3) of Table 7.4 confirm that the coefficient of *Treat*×*Post* remains positive with a significance level of 1%. This finding is consistent with the baseline regression result.

Alternative estimation methods Due to the non-negativity of exploratory innovation values, linear regression model estimates may be subject to bias. To address this issue, this chapter employs the explicit count model of the Tobit method, as suggested by prior research (e.g., Hu et al., 2021a; Van de Vrande, 2013), to confirm the robustness of the aforementioned conclusions. The findings presented in Column (4) of Table 7.4 demonstrate that the estimated effect of *Treat*×*Post* remains significant at the 1% level, thereby corroborating the main results of this chapter.

Placebo test In this section, a placebo test is performed to eliminate the impact of other random factors, in order to obtain a more reliable causal identification effect. Specifically, the placebo test is conducted by randomly selecting the treatment group and repeating 500 repetitions. Then, the t-values of the placebo test results are extracted and plotted in Fig. 7.2. Almost all of the t-values are found to be less than the t-value of the baseline regression. These findings imply that GCP has the potential to stimulate corporate exploratory innovation in green credit-restricted industries. These findings are in complete agreement with the previous conclusions.

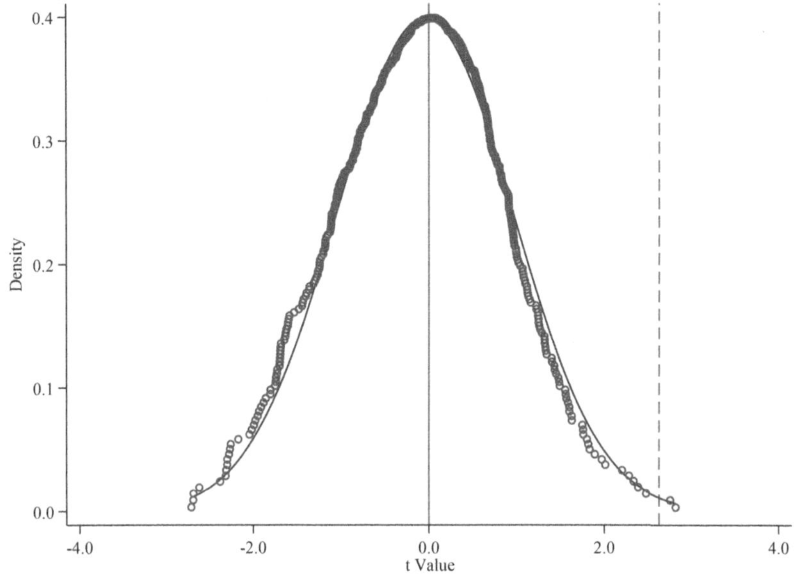

Fig. 7.2 Placebo test. (Notes: The red circles and the curve represent the t-values of *Treat×Post*; the dotted line represents the t-value of *Treat×Post* corresponding to the baseline regression result)

7.5.3 Heterogeneity Analysis

The results of the baseline regression indicate that GCP has a significant impact on promoting corporate exploratory innovation activities. However, it remains unclear whether this effect is heterogeneous across enterprises with varying characteristics, such as ownership, industry, and region. To address this gap, this chapter aims to investigate whether the impact of GCP on exploratory innovation is contingent on the heterogeneous characteristics of ownership, industry, and region.

Ownership heterogeneity The first concern of this chapter pertains to the potential influence of GCP on exploratory innovation, which may be dependent on enterprise ownership. To address this issue, this chapter segregates the entire sample into two categories, namely SOEs and non--SOEs, based on enterprise ownership. Consistent with the methodology

Table 7.5 Heterogeneity analysis results

Variables	*SOEs*	*Non-SOEs*	*High-capital-intensive*	*Low-capital-intensive*	*Eastern*	*Central*	*Western*
	(1)	*(2)*	*(3)*	*(4)*	*(5)*	*(6)*	*(7)*
Treat×post	0.072***	−0.001	0.070***	0.013	0.029	0.094**	0.032
	(3.12)	(−0.04)	(3.25)	(0.45)	(1.30)	(2.41)	(0.72)
Controls	Yes	Yes	Yes	Yes	Yes	Yes	Yes
Enterprise FE	Yes	Yes	Yes	Yes	Yes	Yes	Yes
Year FE	Yes	Yes	Yes	Yes	Yes	Yes	Yes
Observations	5054	6037	6432	4679	7387	2122	1602
Adj. R2	0.285	0.222	0.245	0.247	0.247	0.237	0.258
Chow test *p*-value	0.000***		0.000***		0.000***a		

*** $p \leq 0.01$, ** $p \leq 0.05$, and * $p \leq 0.10$; *t*-statistics are in parentheses. [a] All the *p*-values are significant at the 1% level for eastern versus central, easter versus western, and central versus western

adopted by Zhang et al. (2021), an enterprise is classified as an SOE if its largest shareholder is an SOE or a government agency; otherwise, it is classified as non-SOE. The estimates for the aforementioned categories are presented in Columns (1) and (2) of Table 7.5.

The analysis reveals that the estimate of *Treat×Post* is significantly positive at the 1% level for SOEs, but it is insignificant for non-SOEs at the traditional statistical levels. Additionally, the Chow test is applied to test the difference between SOEs and non-SOEs (*p*-value < 0.01), indicating that there is a significant difference between the two groups at the traditional statistical levels. These outcomes suggest that GCP has a more considerable effect on the exploratory innovation of SOEs compared with non-SOEs, lending support to Hypothesis 7.2. The primary reason for these findings is that SOEs in China mainly predominantly operate in heavily polluting industries, making the effect of GCP on corporate exploratory innovation more pronounced in SOEs (Yao et al., 2021). Specially, on the one hand, SOEs are more inclined to implement the Guideline2012 to achieve technological upgrading in response to national policies. On the other hand, since implementing the Guideline2012, SOEs with low environmental and social risks have had more opportunities to secure long-term and low-cost loans from commercial banks, thereby increasing investment in exploratory innovation activities.

Nevertheless, non-SOEs have limited financing channels and more financing constraints, leading them to allocate their capital to production operations or exploratory innovation activities. Consequently, GCP significantly enhances the exploratory innovation of SOEs in green credit-restricted industries.

Industry heterogeneity The second concern of the heterogeneity analysis aims to investigate whether differences exist in the impact of GCP on corporate exploratory innovation in high- versus low-capital-intensive industries. Nanda and Nicholas (2014) have shown that GCP has a diverse impact on enterprises' innovation. To explore this issue, this chapter separates the whole sample into high- and low-capital-intensive enterprises to explore whether enterprises' different capital requirements drive the influence of GCP on exploratory innovation. The estimates are reported in Columns (3) and (4) of Table 7.5.

The findings reveal that the estimate of *Treat*×*Post* is positive and highly significant at the 1% level. In comparison to high-capital-intensive industries, however, in low-capital-intensive industries it is not significant at the traditional statistical levels. Moreover, the Chow test for coefficients across high- versus low-capital-intensive industries indicates a statistically significant difference at the traditional statistical levels (p-value <0.01). These results support Hypothesis 7.3, suggesting that GCP has a substantial impact on corporate exploratory innovation in high-capital-intensive industries but has no effect in low-capital-intensive industries. This chapter contends that industries creating severe environmental pollution are typically high-capital-intensive industries. These industries require substantial capital to engage in production and operations and undertake R&D activities. Thus, GCP may significantly influence the enterprises' ability to meet their current capital expenditure and future borrowing requirements in high-capital-intensive industries.

Regional heterogeneity The final concern of the heterogeneity analysis aims to determine whether there is a discrepancy in the impact of GCP on corporate exploratory innovation across various regions. As is well-known, there are noticeable variations among regions in terms of marketization, technological innovation, and financing capabilities. Thus, in accordance with the scope of the Regional Development Plan in China, this chapter classifies the sample into eastern, central, and western regions. This is done to explore the distinct impact of GCP on corporate exploratory innovation. The regression is then conducted and presented in Columns (5)–(7) of Table 7.5.

The findings document that the estimate of *Treat×Post* for the central region is significant and positive at the 5% level. Conversely, the estimates for the eastern and western regions are not significant at the traditional statistical levels, which evidences Hypothesis 7.4. Additionally, the Chow test is also conducted on region pairs, and all *p*-values are less than 0.01, indicating that significant structural changes exist among different regions. These results prove that GCP has the potential to significantly promote corporate exploratory innovation in the central regions. Nevertheless, the policy is unlikely to affect corporate exploratory innovation in the eastern and western regions. This can be attributed to the fact that various factors, such as economic development, credit financing policies, marketization degree, environmental regulation, and intellectual property protection, vary widely across different regions of China (Pu et al., 2023; Wang & Lee, 2022). Specifically, the financing channels and innovation capabilities of enterprises in the western region are obviously inadequate. In these circumstances, the limited funds received through GCP may be mainly invested in production and business activities, while relatively little funding is used for R&D activities. Therefore, the penalty effect brought by GCP is far greater than the incentive effect for enterprises in the western region. In contrast, the eastern region has a relatively high degree of marketization, and enterprises have more financing channels. Thus, GCP does not exert significant influence on corporate exploratory innovation in the eastern region.

7.5.4 Underlying Mechanisms

The role of R&D investment The GCP is aimed at promoting the enterprises' transformation and upgrading and enhancing their innovation ability (Hu et al.,2021a; Wen et al., 2023). Green credit-restricted enterprises that meet GCP conditions can obtain the loans needed to increase R&D investment, thus promoting their exploratory innovation activities. Therefore, this chapter explores the influence channel of GCP on corporate exploratory innovation in green credit-restricted industries from the role of R&D investment (*RD*), and constructs the following equation, Eq. (7.2):

$$RD_{it} = a_0 + a\text{Treat}_{it} \times \text{Post}_{it} + \lambda\text{Controls} + \delta_t + \theta_i + \varepsilon_{it} \tag{7.2}$$

Table 7.6 Underlying mechanisms (financing constraints and R& D investment)

Variable	*R&D investment*	*Financing constraints*
	(1)	*(2)*
Treat×post	0.102**	−0.436*
	(2.59)	(−1.93)
Treat×post×FC		0.364**
		(2.15)
FC		−0.793**
		(−2.55)
Controls	Yes	Yes
Enterprise FE	Yes	Yes
Year FE	Yes	Yes
Observations	11,834	11,834
Adj. R2	0.251	0.248

*** $p \leq 0.01$, ** $p \leq 0.05$, and * $p \leq 0.10$; *t*-statistics are in parentheses. The dependent variable in Column (1) is R&D investment, and that in Column (2) is exploratory innovation

where R&D investment (*RD*) is calculated by the natural log of enterprises' R&D expenditure; other variables are the same as in Eq. (7.1).

Column (1) in Table 7.6 presents the estimates on the impact of GCP on R&D investment. The results show that *Treat×Post* has a positive correlation with R&D investment at the 1% significance level. This evidence suggests that GCP can significantly promote enterprises' R&D investment. The importance of R&D investment as a factor for enterprises to carry out exploratory innovation activities is highlighted by this result. That is, GCP promotes corporate exploratory innovation activities by increasing the level of R&D investment, thus supporting Hypothesis 7.6. This finding is consistent with Li et al. (2022), which indicates that GCP enhances innovation efficiency by increasing R&D investment in heavy-polluting industries, because GCP offers financial support to enterprises with low social and environmental risks through special funds and preferential loan interest rates, which reduces the capital cost of R&D investment and encourages enterprises to engage in innovation activities. Moreover, GCP may force enterprises with high environmental and social risks to upgrade by increasing R&D investment to improve their innovation capability.

The role of financing constraints Enterprises need substantial long-term capital investment for exploratory innovation. As such, financing constraints are a crucial factor that affects corporate exploratory innovation (Wang & Wang, 2021). The GCP can amplify the financing threshold and cost for enterprises in green credit-restricted industries (Yao et al., 2021). Concretely speaking, enterprises with low environmental and social risks can more easily secure larger-scale, longer-term, and lower-cost external financing, whereas those with high environmental and social risks encounter challenges in obtaining commercial bank credit (Sharfman & Fernando, 2008; Wang & Wang, 2021). Therefore, this chapter aims to examine whether financing constraints constitute a vital channel through which GCP influences corporate exploratory innovation.

Following Hadlock and Pierce (2010), this chapter adopts the natural log of the absolute value of the *SA* index as the proxy variable of the financing constraints (*FC*), and calculates the *SA* index using the following Eq. (7.3):

$$SA_{it} = -0.737 \times \ln \text{Size}_{it} + 0.043 \times \ln \text{Size}_{it}^2 - 0.04\text{Age}_{it} \tag{7.3}$$

where $\ln\text{Size}_{it}$ is the natural log of the total assets (the unit is million RMB) for enterprise i at year t; Age_{it} is the observation year t of enterprise i minus the establishment year.

To investigate the role of financing constraints in the above relationship, the following Eq. 7.4 is set:

$$\begin{aligned} Y_{it} = \alpha + \beta_1 \text{Treat}_{it} \times \text{Post}_{it} + \beta_2 FC_{it} + \beta_3 \text{Treat}_{it} \\ \times \text{Post}_{it} \times FC_{it} + \lambda \text{Controls} + \delta_t + \theta_i + \varepsilon_{it} \end{aligned} \tag{7.4}$$

where *FC* represents the financing constraints.

The estimated results of the financing constraint mechanism are presented in Column (2) of Table 7.6. The estimates of *Treat×Post×FC* are significant and positive at the traditional levels. These findings indicate that financing constraints play a positive moderating role in the effect of GCP on corporate exploratory innovation in green credit-restricted industries. This suggests that GCP has a more significant impact on corporate exploratory innovation with the increase in the intensity of financing constraints. Furthermore, the estimates of financing constraints (FC) are significant at the 1% or 5% levels, implying that financing constraints impede enterprises from increasing investment in exploratory innovation. These findings suggest that the ability of enterprises to access external credit is a vital factor influencing the impact of GCP on corporate exploratory innovation. The aforementioned conclusion supports Hypothesis 7.5 and is consistent with the study of Luo et al. (2021), which provides evidence that GCP can promote corporate innovation by easing financing constraints, particularly for enterprises with relatively high innovation capabilities that are deeply affected by financing constraints.

One possible explanation for these findings is that GCP can provide long-term and low-cost funds to enterprises with high financing constraints, enabling them to carry out exploratory innovation activities. Some enterprises with low financing constraints can obtain financing resources from various channels, thus having enough financial support to conduct exploratory innovation activities. As a result, GCP has less of an impact on exploratory innovation in green credit-restricted industries. Conversely, the R&D activities of enterprises require significant financial support, and high financing constraints will make it difficult for enterprises to obtain sufficient financial support, thus affecting their exploratory innovation activities. Therefore, GCP can provide long-term and low-cost funds to enterprises with characteristics of low social and environmental risks to carry out exploratory innovation activities.

7.6 Discussions

7.6.1 Contributions

This chapter makes significant contributions in three aspects. Firstly, it explores, for the first time, the impact of GCP on corporate exploratory innovation. The findings highlight that GCP can provide accessible, convenient, and cost-effective credit financing to enterprises with low environmental and social risks. Consequently, these enterprises are more inclined to engage in exploratory innovation activities, which are typically associated with high costs, uncertainty, and risks. Thus, it enriches the existing theoretical research on GCP and corporate exploratory innovation. Secondly, it provides a comprehensive understanding of how China's GCP influences exploratory innovation, taking into account the characteristics of different enterprises. Notably, this chapter identifies the channels through which GCP affects corporate exploratory innovation, including R&D investment and financing constraints. Thirdly, this chapter has important policy implications. It confirms the potential of government-led promotion of corporate exploratory innovation through the implementation of green credit policies. However, it also emphasizes the need for policymakers to consider enterprise types and regional differences when adjusting green credit policies. Only by tailoring policies to local conditions can green policies effectively fulfill their positive role.

7.6.2 Policy Implications

This chapter has significant implications for the implementation and optimization of green credit policies. First, the findings suggest that the government should continue to vigorously optimize green credit policies. This includes expanding the supply of green credit and reducing financing costs, which will encourage enterprises to engage in exploratory innovation activities and promote industrial transformation and upgrading. Additionally, the government should reinforce examination and supervision measures on commercial bank loan behavior. Second, the government should focus more on the heterogeneity of the effects of GCP, and formulate relevant policies to eliminate policy asymmetry. This chapter shows that GCP has a more remarkable influence on the exploratory innovation of SOEs, high-capital-intensive industries, and enterprises in the central region. Therefore, the government should introduce policies to reduce the asymmetric effect of GCP and promote the extension of green credit to low-capital-intensive industries and enterprises in western China. Finally, the government should strengthen the integration of GCP and R&D subsidy policy. The findings indicate that financing constraints and R&D investment are important economic channels through which GCP influences the exploratory innovation of enterprises in green credit-restricted industries. Hence, the government should guide and encourage enterprises that obtain green credit to increase R&D investment and increase credit support to enterprises with considerable financing constraints through incentive mechanisms such as financial interest rates and risk compensation.

There are several significant limitations that should be taken into account for future research. First, this chapter solely investigates the impact of GCP on corporate exploratory innovation, as data on the impact of GCP on green-related exploratory innovation is not available. Therefore, it would be interesting to evaluate this impact in future research. Second, this chapter only considers the heterogeneous effects of ownership, industry, and region. Future studies should explore these effects from other perspectives, such as enterprise scale, green industries, and marketization degree. Finally, this chapter only examined the underlying mechanisms from financing constraints and R&D investment. Therefore, it would be worthwhile to investigate the role of corporate risk-taking and external cooperation capabilities in future works.

7.7 Conclusions

In recent years, the Chinese government has implemented a series of innovation policies with the objective of promoting original innovation and encouraging enterprises to upgrade and transform their industries. These policies have proven to be particularly advantageous for traditional industries operating at low technological levels, as they can leverage innovative approaches to explore new fields, thereby fostering technological progress and promoting enterprise value growth. The issuance of the Guideline2012 has provided guidance to enterprises in green credit-restricted industries on promoting transformation and upgrading through differentiated credit resource allocation policies. This chapter aims to investigate the impact of GCP on exploratory innovation of enterprises in green credit-restricted industries using DID method.

On the basis of the sample of China's listed enterprises from 2007 to 2017, this chapter demonstrates that GCP has a significant impact on corporate exploratory innovation in green credit-restricted industries. This finding is robust and supported by a set of robustness tests. Furthermore, heterogeneity analysis reveals that the impact of GCP on corporate exploratory innovation varies in line with the enterprise characteristics such as ownership, industry, and region. Specifically, SOEs, high-capital-intensive industries, and enterprises in the central region benefit the most from GCP's promotion of exploratory innovation. The underlying mechanisms indicate that R&D investment is the most critical influencing channel of GCP on corporate exploratory innovation. That is, GCP promotes exploratory innovation by increasing R&D investment levels. Moreover, financing constraints positively moderate the impact of GCP on corporate exploratory innovation in green credit-restricted industries. This means that GCP has a more significant impact on exploratory innovation for enterprises facing higher financing constraints.

References

Aghion, P., Askenazy, P., Berman, N., Cette, G., & Eymard, L. (2012). Credit constraints and the cyclicality of R&D investment: Evidence from France. *Journal of the European Economic Association, 10*(5), 1001–1024. https://doi.org/10.1111/j.1542-4774.2012.01093.x

Alexiev, A. S., Jansen, J. J. P., Van den Bosch, F. A. J., & Volberda, H. W. (2010). Top management team advice seeking and exploratory innovation: The moderating role of TMT heterogeneity. *Journal of Management Studies, 47*(7), 1343–1364. https://doi.org/10.1111/j.1467-6486.2010.00919.x

Bai, P., Tang, Y. X., Zhang, W. K., & Zeng, M. (2021a). Does economic policy uncertainty matter for healthcare expenditure in China? A spatial econometric analysis. *Frontiers in Public Health, 9*, 673778. https://doi.org/10.3389/fpubh.2021.673778

Bai, X., Wang, Q. T., Sheng, S. B., & Li, J. J. (2021b). Cross-level interpersonal ties and IJV innovation: Evidence from China. *Journal of Business Research, 134*, 618–630. https://doi.org/10.1016/j.jbusres.2021.05.047

Cheng, C., Mu, J. H., & Liu, G. C. (2025). Do off-Site R&D Institutions Matter for corporate exploratory innovation? *International Journal of Finance & Economics*. https://doi.org/10.1002/ijfe.70020

Chiu, Y. B., & Lee, C. C. (2020). Effects of financial development on energy consumption: The role of country risks. *Energy Economics, 90*, 104833. https://doi.org/10.1016/j.eneco.2020.104833

Czarnitzki, D., Hottenrott, H., & Thorwarth, S. (2011). Industrial research versus development investment: The implications of financial constraints. *Cambridge Journal of Economics, 35*(3), 527–544. https://doi.org/10.1093/cje/beq038

Du, J., Shen, Z., Song, M., & Vardanyan, M. (2023). The role of green financing in facilitating renewable energy transition in China: Perspectives from energy governance, environmental regulation, and market reforms. *Energy Economics, 120*, 106595. https://doi.org/10.1016/j.eneco.2023.106595

Enkel, E., & Gassmann, O. (2010). Creative imitation: Exploring the case of cross-industry innovation. *R & D Management, 40*(3), 256–270. https://doi.org/10.1111/j.1467-9310.2010.00591.x

Gao, Y. C., Hu, Y. M., Liu, X. L., & Zhang, H. R. (2021). Can public R&D subsidy facilitate firms' exploratory innovation? The heterogeneous effects between central and local subsidy programs. *Research Policy, 50*(4), 104221. https://doi.org/10.1016/j.respol.2021.104221

Guan, J., & Liu, N. (2016). Exploitative and exploratory innovations in knowledge network and collaboration network: A patent analysis in the technological field of nano-energy. *Research Policy, 45*(1), 97–112. https://doi.org/10.1016/j.respol.2015.08.002

Hadlock, C. J., & Pierce, J. R. (2010). New evidence on measuring financial constraints: Moving beyond the KZ index. *Review of Financial Studies, 23*(5), 1909–1940. https://doi.org/10.1093/rfs/hhq009

Hu, G. Q., Wang, X. Q., & Wang, Y. (2021a). Can the green credit policy stimulate green innovation in heavily polluting enterprises? Evidence from a quasi-natural experiment in China. *Energy Economics, 98*, 105134. https://doi.org/10.1016/j.eneco.2021.105134

Hu, H., Chen, L. Y., Zhou, Y. X., Liu, Y., & Qu, Q. Z. (2021b). Unpacking the effects of organisational slack on exploitative and exploratory innovation: A study of knowledge-intensive firms. *Innovation-Organization & Management, 25*(1), 57–79. https://doi.org/10.1080/14479338.2021.1989307

Huang, H., Mbanyele, W., Wang, F., Song, M., & Wang, Y. (2022). Climbing the quality ladder of green innovation: Does green finance matter? *Technological Forecasting and Social Change, 184*, 122007. https://doi.org/10.1016/j.techfore.2022.122007

Jia, N., Huang, K. G., & Zhang, C. M. (2019). Public governance, corporate governance, and firm innovation: An examination of state-owned enterprises. *Academy of Management Journal, 62*(1), 220–247. https://doi.org/10.5465/amj.2016.0543

Jiang, F. F., Guo, H., Wei, Z. L., & Wang, D. H. (2018). The fit between managerial ties and resource bundling capabilities: Implications for performance in manufacturing firms. *IEEE Transactions on Engineering Management, 65*(2), 216–226. https://doi.org/10.1109/TEM.2017.2785387

Karamanos, A. G. (2012). Leveraging micro- and macro-structures of embeddedness in alliance networks for exploratory innovation in biotechnology. *R & D Management, 42*(1), 71–89. https://doi.org/10.1111/j.1467-9310.2011.00664.x

Kong, D., Wang, Y., & Zhang, J. (2020). Efficiency wages as gift exchange: Evidence from corporate innovation in China. *Journal of Corporate Finance, 65*, 101725. https://doi.org/10.1016/j.jcorpfin.2020.101725

Kong, D., Zhu, L., & Qin, N. (2022). Does corruption shape firm centralisation? Evidence from state-owned enterprises in China. *Accounting and Finance, 62*(3), 3365–3395. https://doi.org/10.1111/acfi.12890

Lee, C. C., & Wang, F. (2022). How does digital inclusive finance affect carbon intensity? *Economic Analysis and Policy, 75*, 174–190. https://doi.org/10.1016/j.eap.2022.05.010

Lee, C. C., Wang, C. W., & Ho, S. J. (2020). Country governance, corruption, and the likelihood of firms' innovation. *Economic Modelling, 92*, 326–338. https://doi.org/10.1016/j.econmod.2020.01.013

Lee, C. C., Wang, C. W., & Pan, C. M. (2021). Do financial inclusion and R&D matter? Exploring the effect of uncertainty on economic performance. *Applied Economics, 54*(3), 307–325. https://doi.org/10.1080/00036846.2021.1950909

Li, R., & Chen, Y. (2022). The influence of a green credit policy on the transformation and upgrading of heavily polluting enterprises: A diversification perspective. *Economic Analysis and Policy, 74*, 539–552. https://doi.org/10.1016/j.eap.2022.03.009

Li, S., Zhang, W., & Zhao, J. (2022). Does green credit policy promote the green innovation efficiency of heavy polluting industries?—Empirical evidence from China's industries. *Environmental Science and Pollution Research, 29*(31), 46721–46736. https://doi.org/10.1007/s11356-022-19055-8

Liu, S. Y., Du, J., & Zhang, W. K. (2021a). Innovation quantity or quality? The role of political connections. *Emerging Markets Review, 48*, 100819. https://doi.org/10.1016/j.ememar.2021.100819

Liu, S. Y., Du, J., Zhang, W. K., & Tian, X. L. (2021b). Opening the box of subsidies: Which is more effective for innovation? *Eurasian Business Review, 11*(3), 421–449. https://doi.org/10.1007/s40821-020-00178-2

Liu, Y., Li, J., Liu, G., & Lee, C. C. (2021c). Economic policy uncertainty and firm's cash holding in China: The key role of asset reversibility. *Journal of Asian Economics, 74*, 101318. https://doi.org/10.1016/j.asieco.2021.101318

Liu, H. Y., Wang, Y. H., & Cao, J. D. (2025a). Green credit policy and investment-financing maturity mismatch: Evidence from China's green companies. *Environment Development and Sustainability*. https://doi.org/10.1007/s10668-025-06625-2

Liu, M., Yaacob, M. H., Ma, Q. B., & Ding, S. Y. (2025b). Green finance and corporate green innovation: A systematic literature review. *Sage Open, 15*(3), 21582440251370800. https://doi.org/10.1177/21582440251370800

Luo, S., Yu, S. H., & Zhou, G. Y. (2021). Does green credit improve the core competence of commercial banks? Based on quasi-natural experiments in China. *Energy Economics, 100*, 105335. https://doi.org/10.1016/j.eneco.2021.105335

Lv, C. C., Bian, B. C., Lee, C. C., & He, Z. W. (2021a). Regional gap and the trend of green finance development in China. *Energy Economics, 102*, 105476. https://doi.org/10.1016/j.eneco.2021.105476

Lv, C. C., Shao, C. H., & Lee, C. C. (2021b). Green technology innovation and financial development: Do environmental regulation and innovation output matter? *Energy Economics, 98*, 105237. https://doi.org/10.1016/j.eneco.2021.105237

March, J. G. (1991). Exploration and exploitation in organizational learning. *Organization Science, 2*(1), 71–87.

Nanda, R., & Nicholas, T. (2014). Did bank distress stifle innovation during the great depression? *Journal of Financial Economics, 114*(2), 273–292. https://doi.org/10.1016/j.jfineco.2014.07.006

Nemlioglu, I., & Mallick, S. K. (2017). Do managerial practices matter in innovation and firm performance relations? New evidence from the UK. *European Financial Management, 23*(5), 1016–1061. https://doi.org/10.1111/eufm.12123

Nemlioglu, I., & Mallick, S. (2021). Effective innovation via better management of firms: The role of leverage in times of crisis. *Research Policy, 50*(7), 104259. https://doi.org/10.1016/j.respol.2021.104259

Nguyen-Anh, T., Hoang-Duc, C., Nguyen-Thi-Thuy, L., Vu-Tien, V., Nguyen-Dinh, U., & To-The, N. (2022). Do intangible assets stimulate firm performance? Empirical evidence from Vietnamese agriculture, forestry and fishery small-and medium-sized enterprises. *Journal of Innovation & Knowledge, 7*(3), 100194. https://doi.org/10.1016/j.jik.2022.100194

Pang, T. F., Yan, S., & Liu, Z. L. (2025). Does green credit policy inspire enterprise low-carbon technology innovation? *Environment Development and Sustainability*. https://doi.org/10.1007/s10668-025-06399-7

Pu, X. H., Zeng, M., & Zhang, W. K. (2023). Corporate sustainable development driven by high-quality innovation: Does fiscal decentralization really matter? *Economic Analysis and Policy, 78*, 273–289. https://doi.org/10.1016/j.eap.2023.03.008

Shao, H., Wang, Y., Wang, Y., & Li, Y. (2022). Green credit policy and stock price crash risk of heavily polluting enterprises: Evidence from China. *Economic Analysis and Policy, 75*, 271–287. https://doi.org/10.1016/j.eap.2022.05.007

Song, M., Xie, Q., & Shen, Z. (2021). Impact of green credit on high-efficiency utilization of energy in China considering environmental constraints. *Energy Policy, 153*, 112267. https://doi.org/10.1016/j.enpol.2021.112267

Su, Z. F., & Yang, H. B. (2018). Managerial ties and exploratory innovation: An opportunity-motivation-ability perspective. *IEEE Transactions on Engineering Management, 65*(2), 227–238. https://doi.org/10.1109/TEM.2017.2782730

Su, C. W., Li, W., Umar, M., & Lobonţ, O. R. (2022). Can green credit reduce the emissions of pollutants? *Economic Analysis and Policy, 74*, 205–219. https://doi.org/10.1016/j.eap.2022.01.016

Tian, X. L., Kou, G., & Zhang, W. K. (2020). Geographic distance, venture capital and technological performance: Evidence from Chinese enterprises. *Technological Forecasting and Social Change, 158*, 120155. https://doi.org/10.1016/j.techfore.2020.120155

Tingbani, I., Salia, S., Hussain, J. G., & Alhassan, Y. (2021). Environmental tax, SME financing constraint, and innovation: Evidence from OECD countries. *IEEE Transactions on Engineering Management, 70*(3), 1006–1025. https://doi.org/10.1109/TEM.2021.3110812

Van de Vrande, V. (2013). Balancing your technology-sourcing portfolio: How sourcing mode diversity enhances innovative performance. *Strategic Management Journal, 34*(5), 610–621. https://doi.org/10.1002/smj.2031

Voutsinas, I., Tsamadias, C., Carayannis, E., & Staikouras, C. (2018). Does research and development expenditure impact innovation? Theory, policy and practice insights from the Greek experience. *Journal of Technology Transfer, 43*(1), 159–171. https://doi.org/10.1007/s10961-015-9454-3

Wan, G. C., Zhang, W. K., & Li, C. (2024). How does low-carbon city pilot policy catalyze companies toward ESG practices? Evidence from China. *Economic Analysis and Policy, 81*, 1593–1670. https://doi.org/10.1016/j.eap.2024.02.036

Wang, E. Z., & Lee, C. C. (2022). The impact of clean energy consumption on economic growth in China: Is environmental regulation a curse or a blessing? *International Review of Economics & Finance, 77*, 39–58. https://doi.org/10.1016/j.iref.2021.09.008

Wang, X., & Wang, Y. (2021). Green credit policies increase green innovation research. *Journal of Management World, 37*(6), 173–188. (in Chinese).

Wang, C. L., Rodan, S., Fruin, M., & Xu, X. (2014). Knowledge networks, collaboration networks, and exploratory innovation. *Academy of Management Journal, 57*(2), 484–514. https://doi.org/10.5465/amj.2011.0917

Wang, H., Chen, H., Ye, S., & Yin, J. Y. (2024). The impact of green finance on companies' overcapacity: Evidence from green financial reform and innovation policy in China. *Economic Analysis and Policy, 82*, 1320–1336. https://doi.org/10.1016/j.eap.2024.05.016

Wen, H. W., Lee, C. C., & Zhou, F. X. (2021a). Green credit policy, credit allocation efficiency and upgrade of energy-intensive enterprises. *Energy Economics, 94*, 105099. https://doi.org/10.1016/j.eneco.2021.105099

Wen, J. Y., Qualls, W. J., & Zeng, D. M. (2021b). To explore or exploit: The influence of inter-firm R&D network diversity and structural holes on innovation outcomes. *Technovation, 100*, 102178. https://doi.org/10.1016/j.technovation.2020.102178

Wen, H. W., Lee, C. C., & Zhou, F. X. (2022). How does fiscal policy uncertainty affect corporate innovation investment? Evidence from China's new energy industry. *Energy Economics, 105*, 105767. https://doi.org/10.1016/j.eneco.2021.105767

Wen, H., Chen, S., & Lee, C. C. (2023). Impact of low-carbon city construction on financing, investment, and total factor productivity of energy-intensive enterprises. *The Energy Journal, 44*(2), 79–101. https://doi.org/10.5547/01956574.44.2.hwen

Yang, J. Y., Zuo, Z. L., Li, Y. L., & Guo, H. X. (2024). Manufacturing enterprises move towards sustainable development: ESG performance, market-based environmental regulation, and green technological innovation. *Journal of Environmental Management, 372*, 123244. https://doi.org/10.1016/j.jenvman.2024.123244

Yao, S. Y., Pan, Y. Y., Sensoy, A., Uddin, G. S., & Cheng, F. Y. (2021). Green credit policy and firm performance: What we learn from China. *Energy Economics, 101*, 105415. https://doi.org/10.1016/j.eneco.2021.105415

Zeng, M., & Zhang, W. K. (2024). Green finance: The catalyst for artificial intelligence and energy efficiency in Chinese urban sustainable development. *Energy Economics, 139*, 107883. https://doi.org/10.1016/j.eneco.2024.107883

Zhang, Z. G., & Luo, T. Y. (2020). Network capital, exploitative and exploratory innovations—From the perspective of network dynamics. *Technological Forecasting and Social Change, 152*, 119910. https://doi.org/10.1016/j.techfore.2020.119910

Zhang, W. K., & Zeng, M. (2024). Is artificial intelligence a curse or a blessing for enterprise energy intensity? Evidence from China. *Energy Economics, 134*, 107561. https://doi.org/10.1016/j.eneco.2024.107561

Zhang, J., Zhang, W., Song, Q., Li, X., Ye, X. T., Liu, Y., & Xue, Y. W. (2020a). Can energy saving policies drive firm innovation behaviors? Evidence from China. *Technological Forecasting and Social Change, 154*, 119953. https://doi.org/10.1016/j.techfore.2020.119953

Zhang, W. K., Tian, X. L., & Yu, A. (2020b). Is high-speed rail a catalyst for the fourth industrial revolution in China? Story of enhanced technology spillovers from venture capital. *Technological Forecasting and Social Change, 161*, 120286. https://doi.org/10.1016/j.techfore.2020.120286

Zhang, W. K., Zhang, X. Y., Tian, X. L., & Sun, F. W. (2021). Economic policy uncertainty nexus with corporate risk-taking: The role of state ownership and corruption expenditure. *Pacific-Basin Finance Journal, 65*, 101496. https://doi.org/10.1016/j.pacfin.2021.101496

Zhang, W. K., Luo, Q., & Liu, S. Y. (2022). Is government regulation a push for corporate environmental performance? Evidence from China. *Economic Analysis and Policy, 74*, 105–121. https://doi.org/10.1016/j.eap.2022.01.018

Zhang, W. K., Luo, Q., Zhang, Y. F., & Yu, A. (2023). Does green credit policy matter for corporate exploratory innovation? Evidence from Chinese enterprises. *Economic Analysis and Policy, 80*, 820–834. https://doi.org/10.1016/j.eap.2023.09.024

Zhang, W. K., Fan, H. X., & Zeng, M. (2025). Unleashing the power of artificial intelligence: A game changer for urban energy efficiency in China. *Sustainable Cities and Society, 126*, 106372. https://doi.org/10.1016/j.scs.2025.106372

Zheng, S., Zhang, X., & Wang, H. (2023). Green credit policy and the stock price synchronicity of heavily polluting enterprises. *Economic Analysis and Policy, 77*, 251–264. https://doi.org/10.1016/j.eap.2022.11.011

Zhuge, L. Q., Freeman, R. B., & Higgins, M. T. (2020). Regulation and innovation: Examining outcomes in Chinese pollution control policy areas. *Economic Modelling, 89*, 19–31. https://doi.org/10.1016/j.econmod.2019.09.041

PART III

Policy Recommendations

CHAPTER 8

Orchestrating the Transition: Policy Levers for a Technology-Finance Synergy

8.1 Catalyzing Change: Designing Smart Regulations and Incentives

8.1.1 Implementing Low-Carbon-Oriented Policies

Given significant disparities among cities in economic structure, energy consumption, technological capability, and governance, low-carbon transition strategies must avoid a "one-size-fits-all" approach. Instead, differentiated pilot programs should be promoted to generate scalable experiences. Major and resource-based cities should prioritize AI-enabled demonstration projects such as smart energy management and intelligent transport systems. Less-developed and smaller cities should focus on building green industry foundations through digital inclusive finance, policy guarantees, and green credit to attract private investment. Multi-stakeholder collaboration across sectors should be strengthened to integrate emissions reduction efforts. Robust evaluation and dynamic adjustment mechanisms are essential to identify context-specific pathways and scale up successful models.

Concurrently, enhancing corporate ESG performance is crucial for high-quality development. Governments should improve regulations and establish mandatory environmental disclosure to increase transparency. Linking ESG performance to financing conditions can incentivize enterprises to adopt sustainable practices. Professional third-party ESG

W. Zhang et al., *Green Transformation in the Digital Age*,
https://doi.org/10.1007/978-981-95-9121-3_8

evaluation should be supported to enhance market recognition of green companies. Furthermore, central and provincial governments should incorporate low-carbon performance into official evaluations and fiscal transfer criteria. Incentives such as additional funding and expanded autonomy should reward high-achieving cities, while underperformers receive stronger supervision and technical support. Public engagement must also be promoted through awareness campaigns and improved infrastructure to foster a culture of sustainable urban development.

8.1.2 *Formulating Differentiated Financial Support Policies*

In implementing GF policies, full consideration must be given to regional differences in economic development, industrial structure, and resource endowment, to avoid the risk of policy failure resulting from a "one-size-fits-all" approach. In regions with weak foundations for GF—such as resource-dependent cities and less-developed areas—more targeted forms of support should be provided, including the establishment of special green credit quotas, higher fiscal interest subsidies, and the introduction of policy-based guarantee institutions, to address gaps in capital supply, financial services, and technology adoption. For high-pollution and high-energy-consumption industries, entry thresholds and differentiated interest rate policies in GF can be employed to create a mechanism of reverse pressure, encouraging enterprises to increase investment in energy conservation, emissions reduction, and technological upgrading. At the same time, differentiated support measures should facilitate the cross-regional flow of capital and technology, enabling the transfer of GF resources, experiences, and models from advanced regions to less-developed areas, thereby sharing the achievements of green development. Such region-specific policy portfolios can help unlock the driving potential of GF for a balanced, nationwide low-carbon economic transition.

Meanwhile, given the diversity in ownership structure, scale, and industry, policy support for green transformation should be differentiated. For state-owned enterprises, green performance evaluation should be further strengthened to reinforce their role as exemplars in green technology investment. For private SMEs, emphasis should be placed on addressing financing difficulties during transformation through instruments such as credit guarantees and green funds, providing both capital and advisory support. Findings from this study indicate that GF policies have a more pronounced effect on innovation in enterprises located in developed regions and in capital-intensive industries, while the effect is relatively weaker in less-developed areas and among SMEs. Therefore, the government should introduce measures to bridge this gap—for instance, establishing dedicated green transformation funds or interest-subsidized loans for SMEs in less-developed regions—to ensure that policy benefits reach a wider spectrum of enterprises. At the same time, for labor-intensive enterprises that face increased energy consumption risks due to AI adoption, supportive measures should be provided to ease transitional challenges, such as assisting in the procurement of high-efficiency equipment and the training of skilled personnel.

8.1.3 Forging Integrated Finance-Technology Policy for Green Transformation

Green transformation depends not only on financial resources but also on technological breakthroughs and innovation-driven momentum. GF policies should therefore be closely integrated with science and technology innovation policies to form a dual-driving mechanism of capital and technology. Specifically, GF instruments such as green credit and green funds should be aligned with R&D subsidies and tax incentives, enabling enterprises receiving financial support to channel more resources into frontier R&D fields, including energy conservation and carbon reduction, clean energy, and green materials, thereby enhancing their capabilities for "zero-to-one" original innovation. In addition, risk compensation funds and special funds for the commercialization of innovative achievements should be established to reduce the technological and market risks faced by enterprises in green innovation—especially for technology-based SMEs and financing-constrained firms—to prevent the vitality of innovation from being stifled by funding bottlenecks. Cross-departmental coordination should also extend to the development of technical standards, intellectual property protection, and technology dissemination, thereby ensuring that green technology achievements are rapidly transformed into productive forces and creating a closed-loop support system that spans R&D, financing, and industrialization. This will provide continuous technological and capital momentum for the green economy.

8.2 Unlocking Finance: Directing Capital to Sustainable Frontiers

8.2.1 Scaling Up the Provision of Diversified Green Finance

To drive a green economic transition, it is essential to secure a stable, sufficient, and cost-effective source of funding. Governments should continuously improve institutional arrangements for GF, optimize policy instruments such as green loans, green bonds, and green funds, and encourage financial institutions to increase credit support for low-carbon projects and the research, development, and application of green technologies, while ensuring risk remains manageable. At the same time, diversified financing channels should be leveraged to attract private capital, thereby alleviating the supply pressure on individual financial sectors and ensuring stability in the provision of green funds. Regarding financing costs, policy-based interest subsidies, risk-sharing mechanisms, and green insurance can be introduced to reduce financing barriers, enabling SMEs and start-ups to engage in the green transformation. Furthermore, the distribution and use of green loans should be subject to stronger supervision, with robust systems for environmental performance evaluation and continuous project monitoring, to ensure that funds are precisely directed to projects that comply with environmental standards and meet sustainable development requirements. This will prevent "greenwashing" practices and enhance the effectiveness and credibility of GF supply.

8.2.2 Developing Innovative Green Financial Instruments

The development of GF should be deeply integrated with digital transformation, leveraging the connectivity and enabling capabilities of financial technology. On the one hand, technologies such as blockchain, big data, and AI can be employed to improve the design, issuance, and risk management of GF products, to facilitate the introduction of diversified instruments such as green bonds, carbon-related derivatives, and sustainability-linked loans, and to enhance transaction and settlement efficiency. On the other hand, digital inclusive finance platforms can overcome the geographical and scale limitations of traditional financial services, extending access to green investment and financing channels to remote areas and SMEs, thereby enabling broader participation in low-carbon projects and the development of green industrial chains. The integration of digitalization and greening can also enable real-time monitoring and full data traceability across the entire life cycle of green projects, thereby improving information transparency and reducing adverse selection and moral hazard. In parallel, a unified digital information platform for GF should be established to integrate data resources from regulators, financial institutions, and project entities, forming a comprehensive and sustainable digital GF ecosystem.

8.2.3 Securing Efficient and Safe Green Capital Allocation

Establish a special fund for green transformation to support regions with fragile ecosystems and enterprises struggling with transition, offering them policy assurances and interest subsidies to facilitate greener practices. For example, the western regions of China face serious challenges such as forest degradation and soil erosion, which hinder the natural carbon cycle's ability to absorb and neutralize human-induced carbon dioxide emissions. Therefore, it is essential to establish a dedicated green fund to support carbon reduction initiatives in these areas. Meanwhile, an intelligent assessment system for green development should be implemented to evaluate the environmental sustainability of projects and their associated default risks, thereby enabling targeted allocation of green funds and minimizing resource misallocation and inefficiencies. Greenwashing not only results in the inefficient allocation of green funds, but also distorts market pricing mechanisms and triggers unfair competition. Therefore, this intelligent assessment system can help ensure that green capital is directed toward genuinely sustainable sectors by effectively identifying and mitigating greenwashing practices, thereby supporting China's green transformation and advancing the goal of carbon neutrality.

8.3 Driving Disruption: Accelerating Technology Breakthroughs

8.3.1 Promoting Low-Carbon Technology Development

In promoting carbon reduction technologies, equal emphasis should be placed on strict regulation and market-based incentives. On the one hand, strict energy consumption and emission standards should be established, and mandatory measures such as carbon emission trading markets should be improved to compel cities and enterprises to develop low-carbon technologies. On the other hand, fiscal rewards, green certification, and preferential treatment in government procurement should be applied to encourage cities and enterprises to proactively enhance their technologies. Combining mandatory and incentive policies can effectively increase their initiative and creativity in boosting tech investment and advancing technological capabilities. Moreover, governments should encourage interfirm cooperation, enabling companies to share green technologies and management experience through industry alliances and supply chain collaboration, thus improving overall efficiency. For example, large enterprises in high-energy-consuming industries could engage in mergers, acquisitions, or joint ventures with SMEs to establish shared green production process platforms, leveraging economies of scale to lower unit energy consumption. A diversified policy toolkit will ensure that enterprises have both external pressure and internal motivation to achieve sustained green and low-carbon operations.

8.3.2 Deepening the Application of AI to Improve Energy Efficiency

As a core driver in the current wave of technological revolution, AI should be fully leveraged in energy conservation and carbon reduction. Governments should formulate policies to promote the deep integration of AI throughout the production and manufacturing process, encouraging enterprises to adopt AI for process optimization and energy management to achieve simultaneous reductions in consumption and gains in efficiency. The findings of this study indicate that introducing industrial robots and other AI technologies in manufacturing can effectively reduce enterprise energy intensity and operating costs. Accordingly, fiscal subsidies, tax reductions, and other incentives can be provided for procuring and developing energy-efficient AI equipment, as well as for deploying applications such as the industrial internet and intelligent control systems. At the urban level, the construction of smart energy management platforms should be accelerated, with AI applied extensively in the power, transportation, and building sectors to enhance intelligent scheduling and optimize efficiency in energy systems, thereby improving overall urban energy efficiency.

8.3.3 Systematizing Technology-Driven Decarbonization in all Social Domains

It is essential to strengthen the deep integration of technology with the green sector and to promote comprehensive innovation in green technologies across all aspects of production and daily life. Enterprises should be encouraged to accelerate their digital transformation and advance the application of digital technologies throughout their entire production and sales processes. By utilizing AI technology to monitor enterprise energy consumption in real time and develop intelligent energy optimization strategies, energy usage in high-consumption stages can be significantly reduced. Furthermore, through blockchain technology, energy consumption across the entire supply chain can be traced, and the carbon footprint of the product's entire life cycle can be documented, thereby fostering a coordinated effort toward carbon reduction throughout the supply chain. With regard to urban development, cities should reinforce the guiding role of cutting-edge technologies in key areas such as energy, industry, transportation, and construction, while also leveraging digital tools to enhance intelligent urban management and improve the quality of smart living.

8.4 Creating Markets: Leveraging Price Signals to Drive Decarbonization

8.4.1 Stimulating Market-Driven Green Innovation with Strategic Incentives

Enterprises are the primary actors in green technological innovation, and governments should focus on creating a policy environment that encourages bold innovation. First, the green innovation incentive system should be improved by providing greater tax incentives, R&D subsidies, and innovation awards to enterprises engaged in clean energy, energy conservation, and environmental protection, thereby lowering innovation costs and risks. Second, financial institutions should be guided to treat enterprises' green innovation capacity as an important criterion in credit allocation, utilizing tools such as green credit and green equity investment to support high-risk, exploratory innovation projects. Special emphasis should be placed on encouraging and guiding traditional high-pollution industries to enhance "0-to-1" disruptive technology R&D, with targeted policy support to help them break path dependence and achieve green upgrading of processes and products. This government–industry–academia–research collaborative approach will strengthen the independent innovation capacity of green technologies and lay a solid foundation for the long-term sustainable development of industry.

8.4.2 Regulating Market Mechanisms to Ensure Fair Green Competition

A successful green transformation depends on a sound market and institutional environment. Governments should accelerate market-oriented reforms, reduce unnecessary administrative intervention, and ensure that enterprises of all ownership types can compete fairly in green industries. Removing market entry barriers and local protectionism will promote the free flow of production factors, attracting more private capital and innovative forces to the development of green industries. In addition, continuous improvement of the business environment is essential, including simplifying administrative approval procedures, reducing tax and fee burdens for enterprises, and strengthening the legal protection of property rights and intellectual property rights. Enhancing policy transparency and consistency will release market vitality, incentivize enterprises to expand green investment and technological innovation, and amplify the multiplier effect of green transformation in the digital era.

8.4.3 Establishing Modern Market Regulatory Frameworks

The rise of digitalization and intelligent technologies should be accompanied by corresponding institutional safeguards. First, the data governance and intellectual property protection framework for AI should be strengthened, including laws and regulations governing data openness and sharing, cybersecurity and privacy protection, and patent protection. These measures will create a favorable environment for AI innovation and application, while boosting confidence among enterprises and the wider society in adopting AI. Second, the intelligent upgrading and modernization of traditional energy infrastructure should be advanced to provide a solid foundation for AI-empowered green transformation. This includes the digitalization of energy and power systems to enable better integration of AI for energy monitoring, smart grid regulation, and demand forecasting. Coordinated development between energy infrastructure and digital technologies will maximize AI's contribution to energy conservation and emissions reduction, while ensuring effective control over AI's own energy consumption.

8.5 Building Foundations: Shaping Zero-Carbon Physical Networks

8.5.1 *Catalyzing Strategic Investment in Digital Infrastructure*

In pursuit of carbon neutrality goals, DI should be regarded as a strategic lever for promoting low-carbon transition. At present, China's DI development still faces challenges of insufficient scale and regional imbalance, which requires increased investment to enhance overall capacity and optimize spatial allocation. The government should prioritize support for eastern and central regions with relatively advanced digital foundations to further advance the development of new infrastructure such as big data centers and 5G networks, leveraging their first-mover advantages to drive nationwide progress. Meanwhile, greater resource allocation should be directed toward regions with weaker digital foundations to narrow the "digital divide" and unlock DI's nationwide potential for reducing CEI. The empirical evidence from this study demonstrates that DI can significantly reduce urban CEI by facilitating industrial upgrading, reducing resource dependency, and stimulating green innovation. Therefore, DI development should be incorporated as a strategic priority within local green development plans.

8.5.2 *Tailoring Digital Infrastructure to Local Contexts*

Given the varied impacts of AI and digital technologies across regions and city types, policy deployment should be tailored accordingly. In less-developed central and western regions, investment in AI-related infrastructure and talent development should be increased to support the introduction of advanced technologies and foster local innovation, avoiding missed opportunities for energy savings due to inadequate digitalization. In resource-based cities and certain southern cities, attention should be paid to potential energy rebound effects from AI development, with proactive measures to ensure clean energy supply and energy efficiency management capacity to offset any additional consumption. Meanwhile, smaller cities should be encouraged to accelerate industrial clustering and realize economies of scale, using digital transformation to enhance energy utilization efficiency. Governments should provide customized technical support programs for different types of cities to ensure that the process of digital technology empowerment for green development is implemented in a regionally appropriate and steady manner, avoiding new sources of energy waste caused by a uniform "one-size-fits-all" approach.

8.5.3 Integrating Low-Carbon into Comprehensive Urban Planning

As hubs of population, industry, and energy consumption, cities are major sources of greenhouse gas emissions and critical arenas for advancing green transformation. Governments should systematically embed the concept of low-carbon development into the entire process of urban master planning, land-use allocation, infrastructure construction, and public service provision. At the top-level design stage, key indicators such as energy efficiency, CEI control, and ecological carrying capacity should be fully considered. By optimizing the spatial layout of transportation, industry, and residential areas, cities can reduce carbon emissions from long-distance commuting and the clustering of high-energy-consuming industries. Governments should promote the development of green public transportation systems, including the deployment of new-energy buses, rail transit, and bicycle travel, to lower energy consumption and pollution in the transport sector. In the building sector, strict enforcement and gradual enhancement of energy efficiency standards should be implemented, alongside active promotion of green buildings, prefabricated construction, and renewable energy adoption, to improve energy utilization across the entire building lifecycle. In addition, IoT- and big data-based smart city management platforms should be established to enable real-time monitoring and optimal scheduling of urban energy systems for water, electricity, and heat, swiftly identifying and eliminating energy waste. The organic integration of planning, construction, and governance will form a long-term mechanism for low-carbon urban development, guiding cities toward more compact, efficient, green, and livable patterns.

8.6 Strengthening Collaboration: Mobilizing Multi-Stakeholder Synergy

8.6.1 Strengthening Multilevel Government Coordination

Policies related to GF, technological innovation, and industrial transformation need to be coordinated to avoid mutual constraints. Governments should establish cross-departmental coordination mechanisms for green transformation, ensuring unified planning and synchronized implementation of multisectoral policies. Particular attention should be paid to the alignment and balance between new and existing policies—for example, pairing the introduction of stricter environmental standards with necessary transitional fiscal and tax support, so that enterprises are not left uncertain due to policy conflicts. Regular policy evaluations should be conducted, with timely adjustments based on feedback, to ensure that all measures work in concert toward the goal of high-quality green development. Furthermore, policy enforcement capacity must be strengthened by improving performance assessment and accountability systems for all levels of government, ensuring that sound policy measures are effectively implemented.

8.6.2 Advancing Interregional Green Collaboration

Green transformation requires coordinated planning and action across regions, especially given the spillover effects of carbon reduction and energy efficiency improvements—where one city's actions often impact neighboring areas. To avoid fragmented efforts and underestimation of regional synergies, governments should establish mechanisms for intercity cooperation to jointly advance low-carbon development. On the one hand, barriers to the flow of production factors between regions should be removed to facilitate the free movement of green energy, environmental data, and technical talent. On the other hand, regular intercity exchange platforms should be created to share successful experiences in low-carbon technology application and GF practices. In particular, given the cross-regional emission-reduction effects of DI and AI applications, adjacent cities should be encouraged to engage in joint planning and coordinated construction to avoid "digital islands" and resource waste. Regional collaboration can produce "1+1>2" emission-reduction outcomes, while enabling the more efficient dissemination of leading cities' experience nationwide to drive achievement of broader carbon reduction goals.

8.6.3 *Championing Global Green Partnerships for Shared Climate Solutions*

In the digital era, green transformation is a global challenge, and countries should work together on the basis of equality and mutual benefit to promote international cooperation and knowledge exchange. Active participation and leadership in bilateral and multilateral agendas on climate change, GF, and low-carbon transition in smart cities is essential. Countries should share effective practices in green credit, digital inclusive finance, and low-carbon technology application, while promoting cross-border technical assistance and capacity building. Through such cooperation, countries can draw on each other's experiences to identify green development pathways suited to their own national contexts, thereby enhancing their capacity to address environmental challenges. At the same time, collaboration and connectivity within global green industrial and supply chains should be deepened, promoting the wider adoption of green technologies and products. This will enable economic development and environmental protection to reinforce each other, advancing global sustainable development goals.

Index[1]

A

B

C

[1] Note: Page numbers followed by 'n' refer to notes.

W. Zhang et al., *Green Transformation in the Digital Age*,
https://doi.org/10.1007/978-981-95-9121-3

The manufacturer's authorised representative in the EU is Springer Nature Customer Service Centre GmbH, Europaplatz 3, 69115 Heidelberg, Germany. If you have any concerns regarding our products, please contact ProductSafety@springernature.com

Printed and bound by CPI Group (UK) Ltd, Croydon, CR0 4YY

21/09/2026

02219945-0007